Fusion 2

Sam Holyman
Phil Routledge
David Sang

Series Editor: Lawrie Ryan

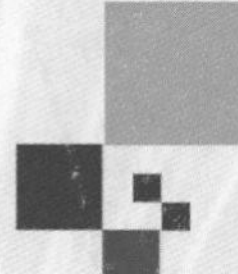

Nelson Thornes

Acknowledgements

The Pupil Book authors wish to express their special thanks to the following people

Neil Roscoe
John Payne
Ruth Miller
Jane Taylor
Judy Ryan
Paul Lister
Nick Pollock
Darren Forbes
Geoff Carr
Bev Routledge
James Routledge
Joel Routledge
Jack Routledge
Sarah Ryan
Amanda Wilson
Annie Hamblin
Doug Thompson

Published in 2008 by:
Nelson Thornes Ltd
Delta Place
27 Bath Road
CHELTENHAM
GL53 7TH
United Kingdom

11 12 / 10 9 8 7 6 5 4 3 2

A catalogue record for this book is available from the British Library

ISBN 978 0 7487 9836 0

Illustrations by GreenGate Publishing, Barking Dog Art, Harry Venning and Roger Penwill

Cover photograph: Photo Library

Page make-up by GreenGate Publishing Services, Tonbridge, Kent

Printed in China by 1010 Printing International Ltd

Contents

Introduction

What is this book about?

By the time you read this you will have found out lots of exciting things from Fusion 1. Just like Fusion 1, this book aims to give **you** the ideas and knowledge to challenge the world around you. There are lots of fun and interesting practical ideas included here to keep you switched on to science.

Building on the skills and knowledge you learnt in Fusion 1, this book will help you to see the connections between the different areas of science. It will also help you to use the skills you have learnt and apply them to new situations.

Discovering how our planet changes over time, understanding the importance of our surroundings and seeing how our planet fits into the universe around us are just some of the things tackled in this book. Should I walk to school today? Should I stop eating meat? Can I do anything to help prevent acid rain? The knowledge you gain here will help you make choices about how you live your life and make everyday decisions.

One of the most amazing things about science is that it gives you the ability to make your own discoveries about the world around you. Fusion 2 will provide you with the skills to be a real scientist, investigating real problems or issues and making your own conclusions based on the evidence and research you collect. It will give you a chance to work individually and as part of a team. At the end you can then present your key findings to your audience. This doesn't just mean through PowerPoint – there are a whole heap of ways to present your scientific discovery!

Just like Fusion 1, there are lots of features in this book to help you. Examples of these are shown on the left and overleaf.

Next time you...

... run to catch the bus, make toast, see lightning, you can link your science knowledge to what you are seeing or doing. It's a great way to remind yourself of what you know.

Link up to...

Another subject. It could be ICT, Maths or Art and Design or any others that you study. Lots of subjects have ideas and skills that overlap.

Stretch Yourself

Some bits of science are easier to understand than others. If you find an area of science easy you might want to stretch yourself by trying some harder tasks.

Questions in the yellow boxes. These are quick questions to check that you understand the science you are learning.

Did You Know?

There are lots of amazing fact's about Science. They are a great way to dazzle friends and family with your fantastic knowledge! Science certainly has the wow factor and sometimes it can be just plain gruesome.

Great Debates

Do you think you're right about everything? Do you enjoy getting into a heated discussion about day to day things, perhaps in the news or about your favorite music group? Being able to discuss and present an argument in science is an important skill. It's also a great way to find out more about interesting scientific issues.

science@work

Science is all around us in our everyday lives and in many people's jobs. Some jobs use science more obviously than others.

activity

Activity

Science is a very practical subject and we have tried to make this book as active as possible. The activities in this book vary from full practical investigations to much smaller activities. They help to develop your knowledge and skills in as fun and active a way as possible.

Some bits of science are difficult to understand and you may need a bit of help. This feature gives you tips and shows you connections that may make it all easier to understand.

Summary Questions

These check that you understand everything. Some questions are easy to answer. Others are more challenging and you may need to ask for help. You can use questions to help see what you understand well and to see where you can improve.

KEY WORDS

Important scientific words are shown in bold and appear in a list on the page and in the glossary at the back of the book.

Body Systems

Lots of cells

You have already learned that living things are made of cells. Some organisms are made of only one cell.

a Give some examples of living things that are made of single cells.

b Can you think of an example of a 'living organism' that is not made of cells? (Hint: Some scientists do not count them as alive!)

Humans are made of millions of cells. Most of the cells in your body are 'specialised'. That means they have a special shape so that they are adapted to carry out a certain job.

Look at these examples of some specialised cells.

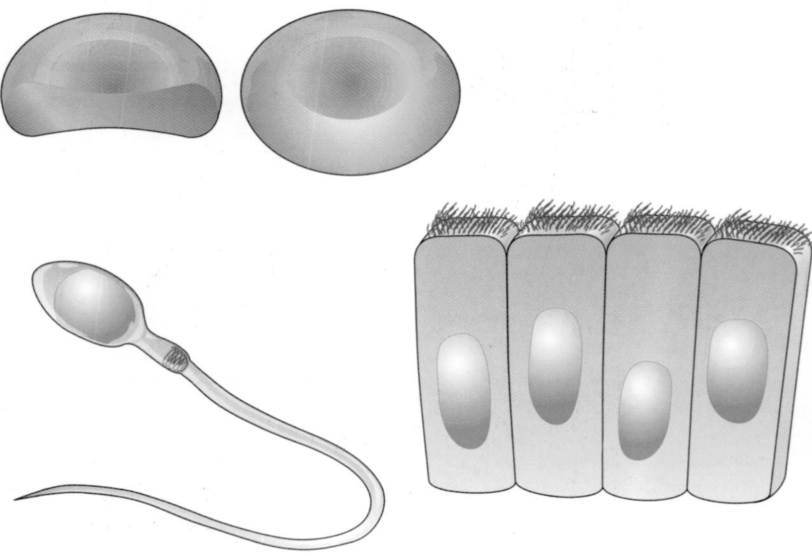

c Can you identify the type of cell shown in each of the diagrams? For each type of cell, describe its job and explain how it is adapted to carry this out.

A group of cells of the same type make a tissue.

Various different types of tissue make up an organ. Organs carry out major life functions such as reproduction, breathing, digestion, etc.

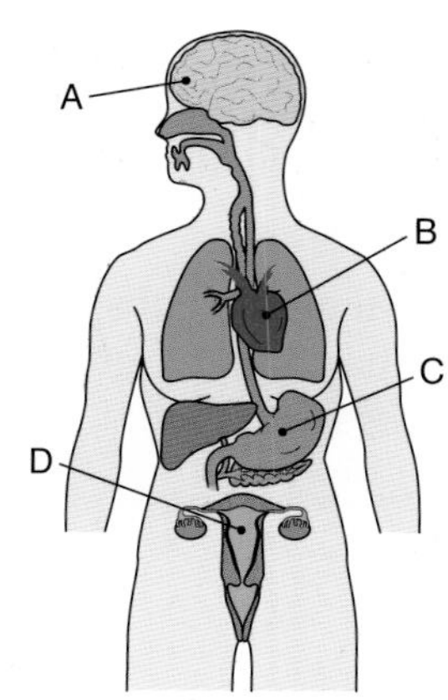

d The diagram shows some organs in the body. Identify the organs A to D. Write a sentence to describe the job of each organ A to D.

e Make a list of the organs connected with feeding and digestion.

You will have noticed that getting nutrients from your food involves quite a lot of organs! Together these make up the digestive system. Your body is made up of several different organ systems.

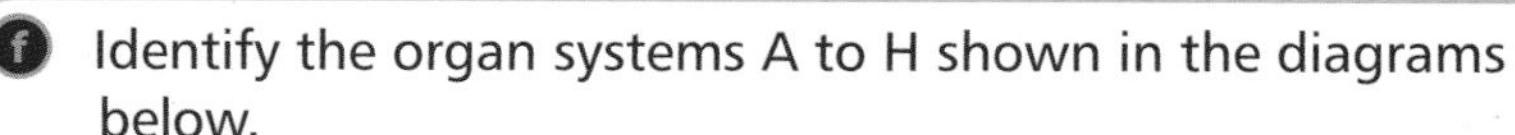

f Identify the organ systems A to H shown in the diagrams below.

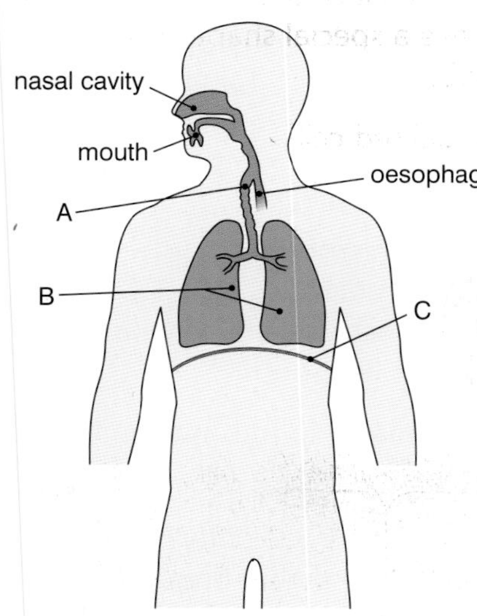

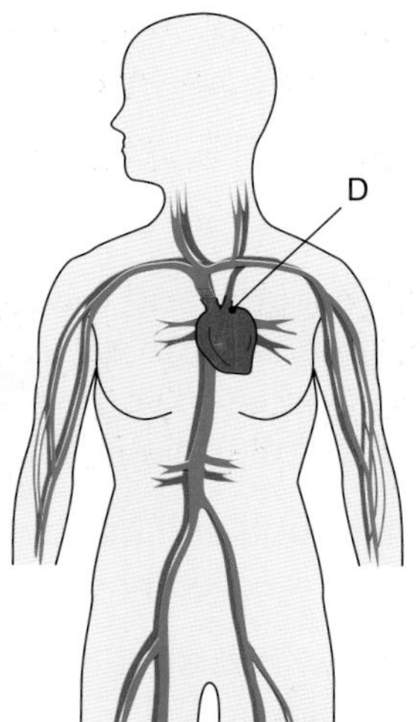

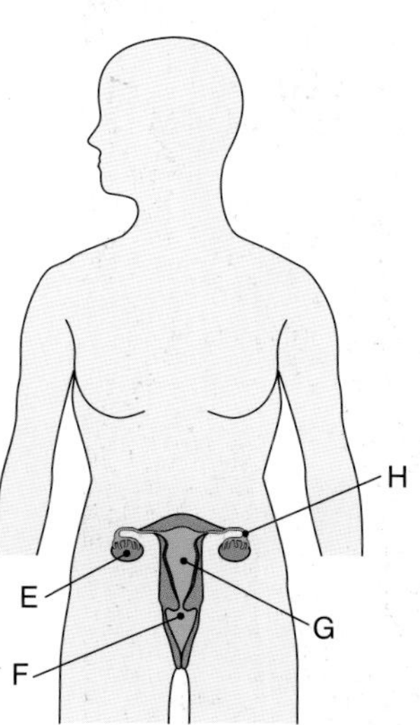

activity

Drawing organs

Lie on the floor and get a partner to draw around your body onto a large sheet of paper (or several sheets of paper joined together).

Draw your body organs on the outline. For each organ, make a label with its name and function.

You could make this into a game where your classmates have to join the name to the organ, then the function to the name.

The Digestive System

B1.2

- What are the different parts of the digestive system?
- What are the jobs of the parts of the digestive system?

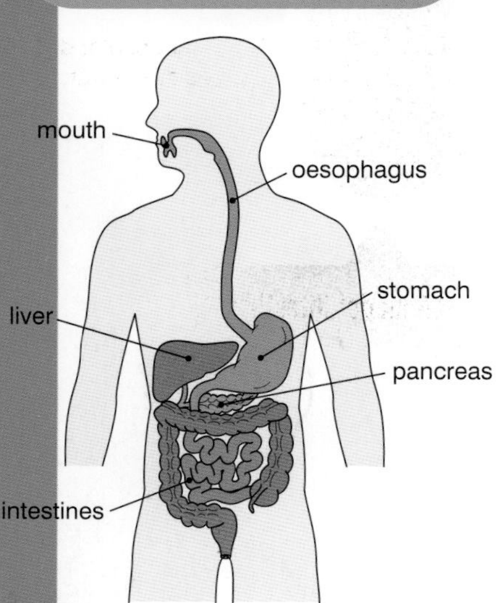

Let's eat!

We need to absorb nutrients from food. Then they can be carried to the parts of the body where they are needed. Starch, fat and protein molecules are too big to pass through the wall of your intestine. One of the jobs of the digestive system is to break down these molecules. This makes them small enough to be absorbed.

Digesting food

The first stage of breaking down food is **chewing**. This makes the pieces of food smaller and mixes them with **saliva**. Saliva makes it easier to swallow the food by making it slippery. It also contains chemicals, called **enzymes**, which break down starch molecules.

a Why is saliva added to food?

When you swallow the food is pushed down the oesophagus to the stomach. Muscles in the wall of the oesophagus squeeze the food along. This process is called **peristalsis**. It's a bit like squeezing toothpaste from the tube!

Food stays in your stomach for a few hours.

- Hydrochloric acid is added to kill any microbes.
- Enzymes start to break down protein molecules.
- The muscles in your stomach wall squash and churn your food so it turns into a mixture looking a bit like soup!

Chewing food and mashing it up into smaller lumps in your intestines is called **physical digestion**.

After a few hours food passes into your small intestine. More enzymes are added from your pancreas. These help break down starch, fat and protein.

Your **liver** makes a substance called **bile**. This passes into your intestine from the **bile duct**. Bile helps the digestive process. It makes it easier for the enzymes to work by increasing the surface area of the fats.

b What is peristalsis?

Did You Know?

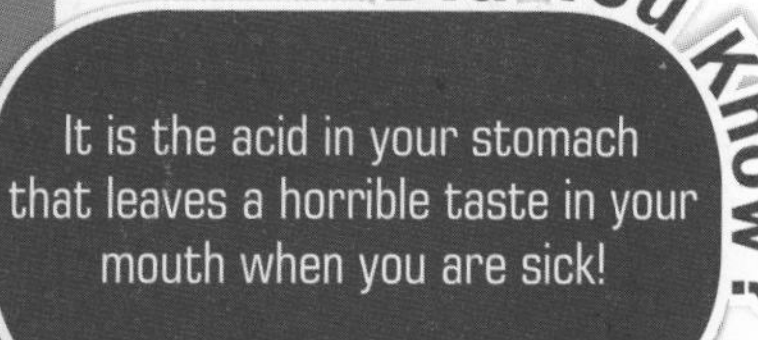

It is the acid in your stomach that leaves a horrible taste in your mouth when you are sick!

Absorbing nutrients

When the molecules in the small intestine have been digested, they are small enough to pass through tiny holes in the intestine wall and into your blood.

The small intestine is lined with tiny finger-like structures called **villi**. These increase the surface area of the intestine so that nutrients can be absorbed quickly.

Villi in the small intestine

c What are villi? Why are they important?

activity

A model intestine

Starch is made of large molecules, while sugar molecules are much smaller. Visking tube has tiny holes, similar to the intestine wall.

Note: To test for starch, add a few drops of iodine solution. If it goes blue-black, starch is present.

To test for sugar, add a few drops of Benedict's solution then heat the mixture. If it goes yellow or orange, sugar is present.

- Fill a length of Visking tubing with a mixture of starch and sugar solution.
- Put the Visking tubing in a boiling tube of warm distilled water.
- After 20 minutes take some water from the boiling tube.
- Test this water to find out if it contains sugar or starch.
- Draw a table to show your results.

The Visking tubing was a model intestine. What part of the body was represented by the water in the boiling tube?

How do the size of sugar and starch molecules compare with the size of the holes in the Visking tubing?

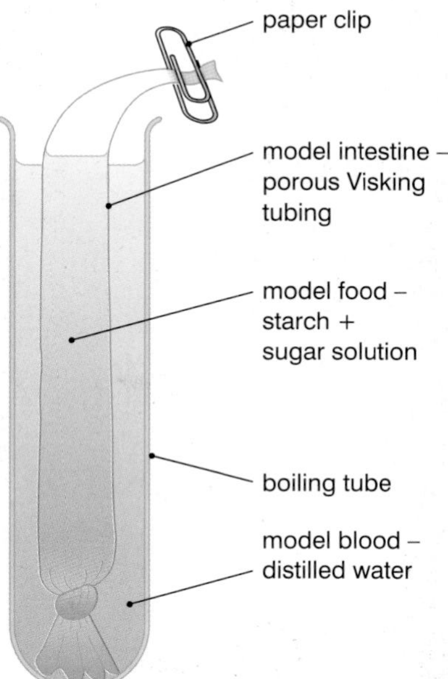

A model intestine

Getting rid of waste

Plant cell walls are made of a substance called cellulose. This cannot be digested. It passes into your large intestine along with millions of bacteria. Excess water is reabsorbed from undigested food in your large intestine. This leaves **faeces** (or poo) which is stored in your rectum. When you go to the toilet faeces is pushed out through your anus.

Fibre is made of cellulose. Although we cannot digest fibre, it helps food to move along the intestine. Without fibre you would become constipated and could develop bowel cancer.

Summary Questions

1. Draw a table to summarise the jobs of these parts of the digestive system: mouth, stomach, small intestine, large intestine, rectum, anus.
2. Explain what fibre is made of and why it is important.

KEY WORDS

chewing
saliva
enzyme
peristalsis
physical digestion
liver
bile
bile duct
villi
faeces

Digesting Food

- How do we digest food molecules?
- How are large molecules such as starch, protein and fat broken down?

Digestion and enzymes

Starch and protein are made of many smaller molecules joined together in long chains. Starch and protein are too big to pass through the wall of your intestine. Your body must break down these molecules so that they are small enough to be absorbed into your blood.

Breaking down large molecules into smaller ones is called **chemical digestion**. **Enzymes** are biological catalysts, made of protein.

 What is an enzyme?

Digestive enzymes speed up the breakdown of large molecules. Different enzymes break down different molecules.

Starch

Starch molecules are made of chains of hundreds of glucose molecules. Starch is broken down into smaller molecules by an enzyme called **amylase**.

Amylase breaks down starch

amylase 'cuts' the starch molecule into separate molecules

Amylase is a type of **carbohydrase** because it breaks down carbohydrate molecules. We produce amylase in our mouths in saliva. If you chew a piece of bread for a few minutes without swallowing, you will notice that it gets sweeter. That's because you are breaking down the starch into sugar. We also find carbohydrases in our intestines.

 Does sugar need to be digested? Explain your answer.

activity

Digesting starch

- Put 10 cm^3 starch solution in a beaker and warm it to 37°C.
- Add 2 cm^3 amylase solution.
- Every two minutes take out some of the mixture and test it with iodine.
- Why is the mixture warmed to 37°C?
- What happened to the mixture?
- How could you check that the starch has been broken down to sugar?

Protein

Protein molecules are made of long chains of smaller molecules called **amino acids**. There are 20 different amino acids. Different proteins are made of different combinations of amino acids. We can make some amino acids ourselves but some we can only get from food. These are called essential amino acids.

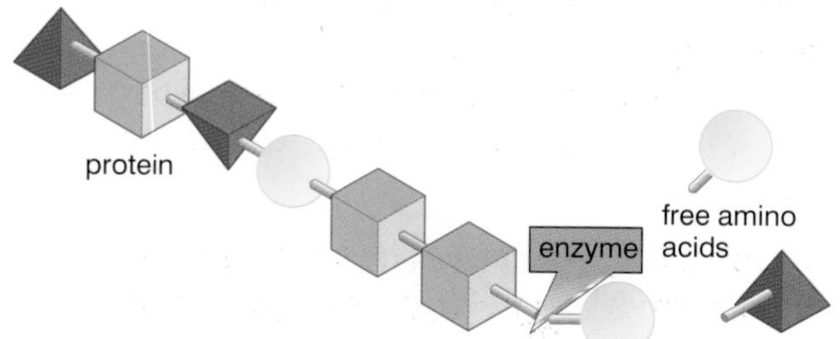

Digesting protein to get amino acids

Link up to...

FOOD TECHNOLOGY

You will have learned about the main food groups in food technology. Make a poster to show which foods are good sources of carbohydrates, proteins and fats.

Enzymes which break down protein are called **proteases**. Proteases are made in gastric juice in your stomach and in the small intestine.

C What do we call the smaller molecules which proteins are made from?

Lipids

Lipid is the scientific name for fat. Enzymes which break down lipids are called **lipases**. We find lipases in our intestines. They break down lipids to form fatty acids and glycerol.

Absorption

When large molecules have been digested they are absorbed into blood vessels inside the villi.

Your blood carries nutrients to the liver. The liver uses some nutrient molecules to give us energy. Other molecules are carried around the body in the blood. Some useful nutrients like vitamin B and iron are stored in the liver. Some glucose is converted into a substance called **glycogen**. This is used as a store of energy in the liver and muscles. Sometimes you eat too much food. Then what you don't use, you turn into fat which is stored around your body.

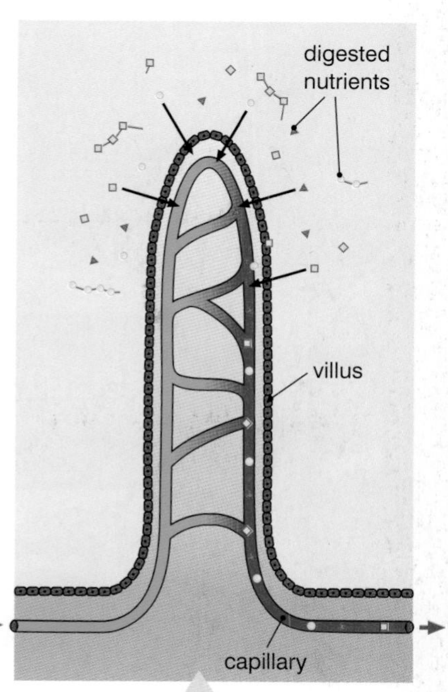

Absorbing nutrients

Summary Questions

1. Complete the table below:

Type of molecule	Where it is digested	Type of enzyme	Product of digestion
starch	mouth and small intestine		
protein		protease	
lipid			fatty acid and glycerol

2. Explain the difference between physical digestion and chemical digestion.

KEY WORDS

chemical digestion
enzyme
amylase
carbohydrate
amino acids
protease
lipase
glycogen

Food and a Balanced Diet

- What are the six food groups?
- How can we identify these in a sample of food?

Did You Know?

In 1979 an Austrian man was arrested by the police and locked in a cell. They then forgot about him for 18 days! He managed to survive without food and only had rain water which he collected through a broken window.

Nutrients

You already know that food contains **proteins**, **carbohydrates** and **fats**. Plant foods also contain fibre. We do not digest fibre but it helps food to move through our digestive system.

Protein

Protein is used:

- to make new cells and for growth
- to heal wounds
- to repair damaged parts of the body.

Hair, fingernails and muscle are made of protein. Protein is also used to make enzymes, antibodies and hormones.

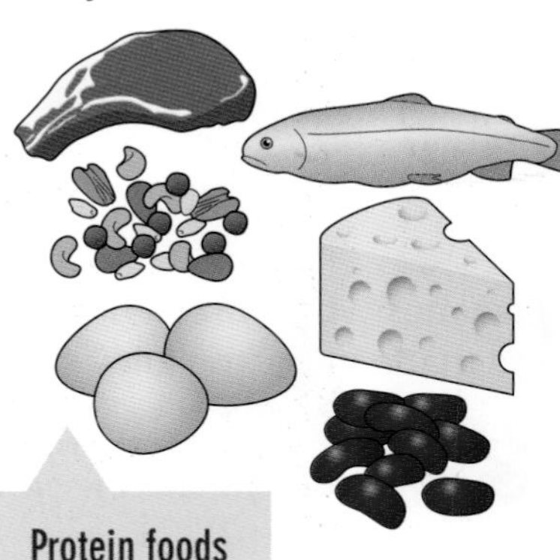

Protein foods

a Why do young people need protein more than adults?

b What is the job of antibodies?

Carbohydrates

Carbohydrates

There are two types of carbohydrate: **sugars** and **starch**. They are our main source of energy. Starchy food is better for us than sugary food as the energy is released more slowly as we need it.

Your body stores unused carbohydrate in the form of fat. Potatoes, rice, pasta, cereals and bread are good sources of starch. Cakes, biscuits, sweets, jam and fizzy drinks contain a lot of sugar.

c Why are starchy foods better for you than sugary foods?

Fats

Fats

Fats contain a lot of energy – about twice as much as the same amount of sugar. Too much fat in your diet makes you overweight, but small amounts of fat are essential. Your body needs fat to make cell membranes and nerve cells.

d Your friend is overweight and decides to cut out all foods that contain fat. What advice would you give?

Minerals and vitamins

Food also contains **vitamins** and **minerals**. These are an essential part of our diet, even though we only need them in very small amounts. People who do not get enough of a particular vitamin will suffer from a deficiency disease.

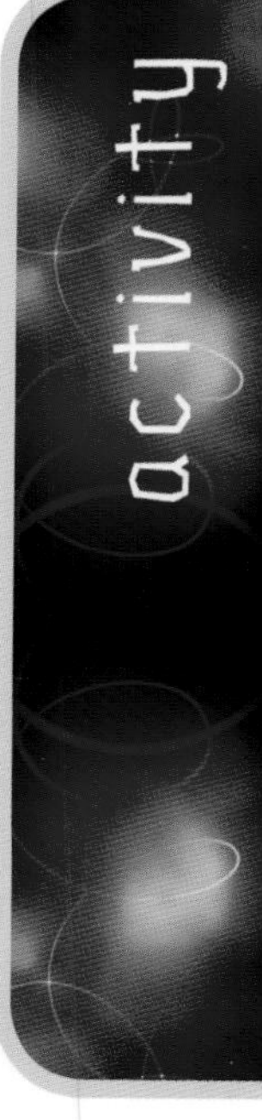

Testing foods

We can use chemical tests to find out if foods contain starch, sugar, protein and fat.

- *Starch test* – Add iodine to the food. If it changes colour from orange to blue-black then the food contains starch.
- *Sugar test* – Add Benedict's solution to the food and heat it. If it contains sugar it changes colour from blue to green, yellow, orange or red. The redder the colour the more sugar it contains.
- *Protein test* – Put the food in a test tube and add about 1 cm^3 of sodium hydroxide solution. Then add drops of copper sulfate solution. (Or add 1 cm^3 of Biuret reagent.) If you see a purple colour the food contains protein.
- *Fat test* – Rub a piece of the food on a filter paper. Hold the paper up to the light. If it goes translucent (allows light to pass through) the food contains fat.

Safety: Wear chemical splash-proof eye protection. Sodium hydroxide is corrosive.

Make a table to show which nutrients were found in each of the foods you tested.

Nutrient	Vitamin A	Vitamin B1	Vitamin C	Vitamin D	Iron	Calcium
How it is used by the body	helping us to see, especially in the dark	making nerves work properly and helping release energy	used by cells	making healthy bones	making red blood cells	making strong bones and teeth
Deficiency disease	night blindness	beriberi	scurvy	rickets	anaemia	rickets
Good sources of the nutrient	dairy products, carrots, green vegetables, oily fish	liver, meat, eggs, cereals, spinach, cabbage, cocoa	oranges, lemons, blackcurrants, kiwi fruit	milk, cheese, eggs, butter, oily fish, made in the skin in sunlight	liver, nuts, wholegrain cereal, eggs, beans	cheese, milk, spinach, sardines and butter

Summary Questions

1. Make a table to summarise what you have learned about the major nutrients:

Nutrient	How the body uses the nutrient	How to test for the nutrient	Foods which are a good source of the nutrient
Sugar			
Starch			
Protein			
Fat			

2. Why does a child need more calcium than an adult?
3. Why does a 20 year old girl need more iron than a 20 year old boy?
4. Find out about James Lind, the naval doctor who discovered how to prevent scurvy.

Link up to...

HEALTH EDUCATION

You will learn about the importance of a balanced diet in health education lessons.

KEY WORDS

protein
carbohydrate
fat
sugar
starch
deficiency disease
vitamin
mineral

Breathing

B1.5

- How do our lungs work?
- How do we breathe in and out?

When you breathe

Breathe in and out. Put your hands on your chest. Observe carefully what is happening to your chest:

- As you breathe out, your chest moves downwards and inwards.
- As you breathe in, your chest moves upwards and outwards.

Breathing out is called **exhaling**. Breathing in is called **inhaling**.

a What does inhaling mean?

b What does exhaling mean?

Now, Sir, if you would like to use your intercostal muscles to increase your thorax volume.
Eh?
Breathe in.

What happens to make you exhale and inhale?

Your lungs are in the upper part of your body called the **thorax**.

- To breathe out, you reduce the volume of your thorax. This increases the pressure of the air in your lungs. The air gets pushed out.
- To breathe in, you increase the volume of your thorax. This decreases the pressure of the air in the lungs. The higher pressure air outside gets pushed into your lungs.

How do you change the size of your thorax?

Below your lungs there is a dome-shaped sheet of muscle called the **diaphragm**. This separates your thorax from the lower part of your body called the **abdomen**.

When the diaphragm contracts it flattens out. This makes the thorax bigger. When the diaphragm relaxes it returns to a dome shape, making the thorax smaller.

When you are relaxed you only breathe in and out about half a litre of air each time. Gentle breathing is due to the movement of the diaphragm.

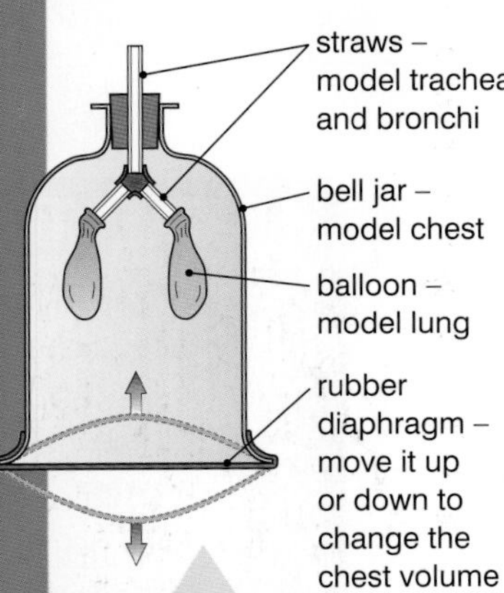

A model chest

activity

How we breathe

Your teacher will show you a model of the chest (thorax). It shows movement of the diaphragm making air move in or out of the lungs.

If you exercise you need to take in more air with each breath. This calls for another set of muscles, found between your ribs. When these **intercostal muscles** contract they lift up your rib cage. This makes your thorax bigger so you breathe in. When these muscles relax your rib cage drops down and air is forced out of your lungs.

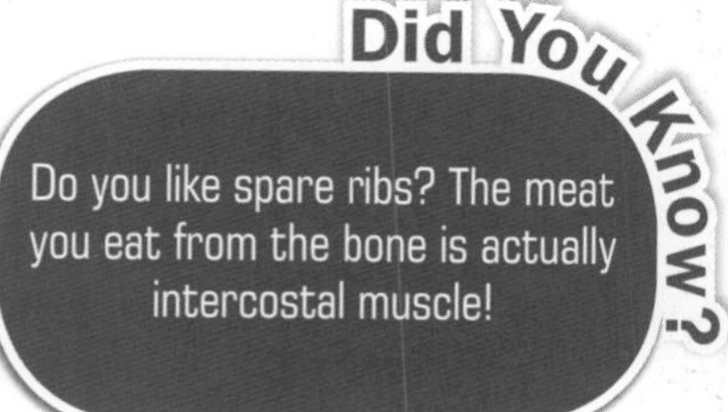

activity

Lung capacity

You will probably do this as a class experiment.

- Mark a large plastic container in litres and half litres. A five-litre fruit juice container is ideal.
- Fill it with water.
- Carefully turn it upside down in a large bowl of water.
- Take a deep breath and blow out through a rubber tube into the plastic container.
- This measures your lung capacity.
- Draw a table of your results or use a spreadsheet.

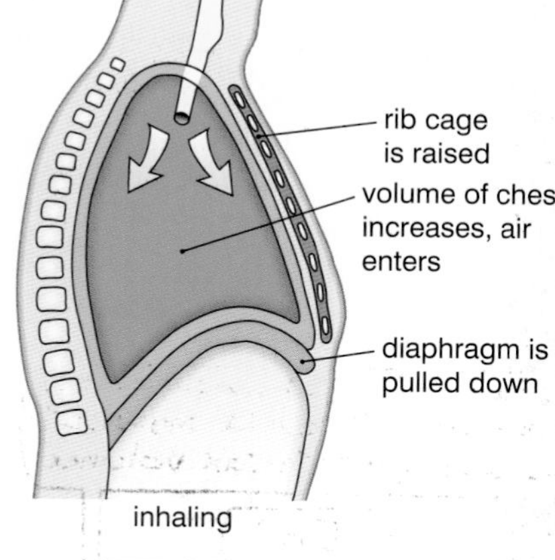

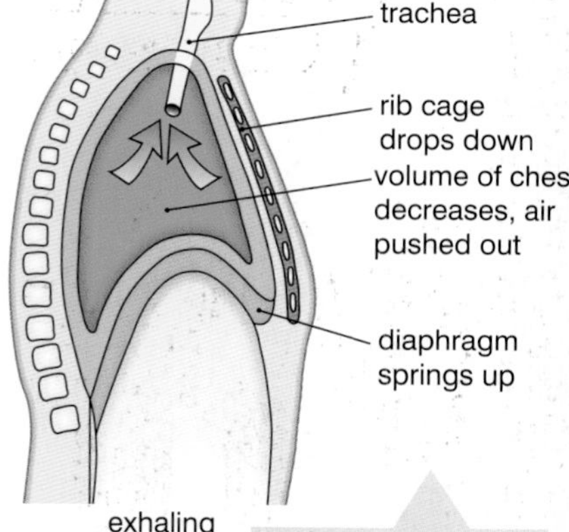

How we breathe

Normally you breathe out by relaxing the diaphragm and intercostal muscles. This makes the thorax get smaller and air is forced out. If you need to blow out very hard you have another set of intercostal muscles. Contracting these muscles forces your rib cage to become smaller. Coughing, sneezing and playing the bagpipes use these muscles!

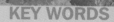

Summary Questions

1. Copy and complete the table to summarise what happens when we breathe in and out:

	What happens to the diaphragm	What happens to the rib cage	What happens to the size of the thorax	How the pressure changes
Breathing in (inhaling)				
Breathing out (exhaling)				

2. What are your thorax and abdomen? What is the name of the muscle that separates them?

KEY WORDS

exhale
inhale
thorax
diaphragm
abdomen
intercostal muscles

Lungs

B1.6

- How are our lungs adapted for gaseous exchange?
- What parts make up the lungs?
- How do our lungs keep themselves clean?

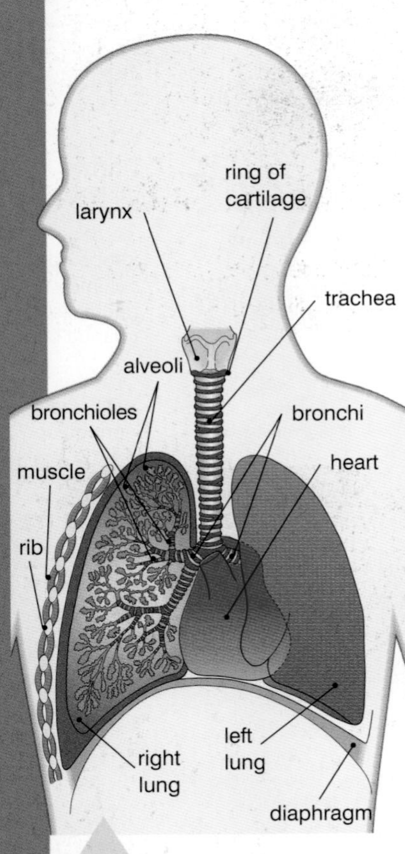

Human lungs

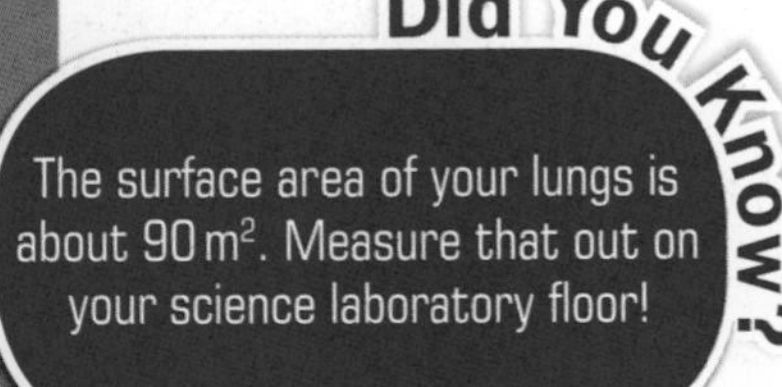

Did You Know?

The surface area of your lungs is about 90 m^2. Measure that out on your science laboratory floor!

All about lungs

Your lungs are in your thorax, protected by your rib cage. Your windpipe, or **trachea**, carries air from the back of your throat to your lungs. Your voice box, or **larynx**, is at the top of your windpipe. This contains your vocal cords.

The bottom of your windpipe has two branches called **bronchi** (singular: 'bronchus'), one to each lung. The trachea and bronchi are surrounded by rings of cartilage. This stops them from being squashed.

The bronchi branch into smaller and smaller tubes called **bronchioles**. At the end of each bronchiole there is an air sac called an **alveolus** (plural: 'alveoli').

a Why are the trachea and bronchi surrounded by rings of cartilage?

Inside the lungs

The alveoli are where gases pass in and out of the lungs. There are about 300 million alveoli in our lungs. This means that there is a large surface area for **gaseous exchange**. Alveoli have very thin walls and each one is surrounded by tiny blood vessels called capillaries.

Inside the alveoli:

- Oxygen **diffuses** from the air into the blood.
- Carbon dioxide diffuses from the blood into the air.

Oxygen passes into the red blood cells. It combines with haemoglobin which carries oxygen around the body.

If there is too much carbon dioxide in your blood it becomes acidic. Special cells in arteries in your neck measure the pH of your blood. If it becomes too acidic, they increase your breathing rate.

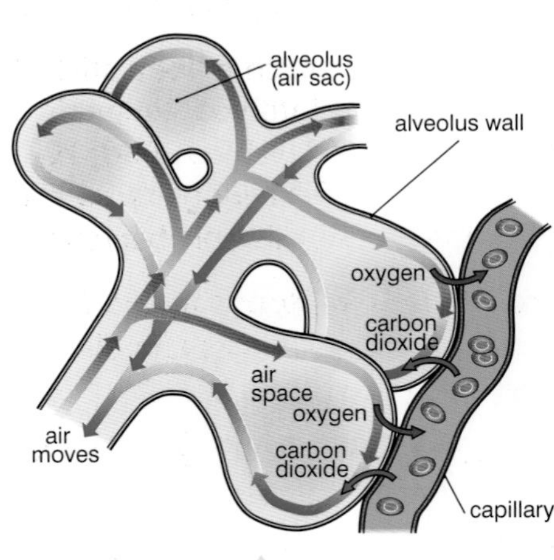

Gas exchange at an alveolus

b Which gas i) passes from the air to your blood and ii) passes from the blood to the air?

 How is your rate of breathing controlled?

Emphysema

Many people who smoke develop **emphysema**. The walls of the alveoli break down so the lungs have a smaller surface area. Also the capillaries around the alveoli are destroyed.

People with emphysema have difficulty in getting enough oxygen. They have to breathe extra oxygen from a cylinder.

 Explain why people who suffer from emphysema are usually short of breath.

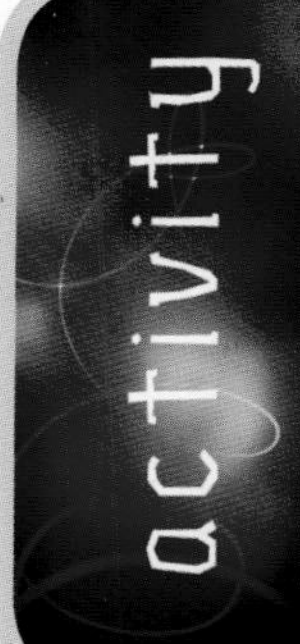

Looking at lungs

Your teacher may show you a set of animal lungs.

- Watch as your teacher inflates the lungs.
- Feel the lungs and describe what they feel like.
- Why do butchers call lungs 'lights'?

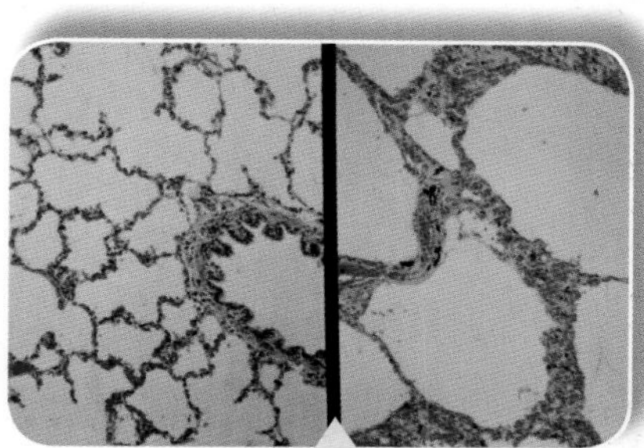

Normal lung (left), lung with emphysema (right)

Keeping the lungs clean

The surface of the bronchioles is made up of two types of cell:

- Goblet cells which make a sticky liquid called **mucus**. If microbes and dust get into your lungs they stick to the mucus.
- Ciliated cells which have tiny hairs called **cilia**. Cilia sweep the mucus out of the lungs up to your throat where it is swallowed. Stomach acid kills only bacteria.

Cigarette smoke stops the cilia from working. Smokers have to cough to remove mucus and dirt from their lungs.

Link up to...

HEALTH EDUCATION

You will learn more about the effects of smoking in health education lessons.

Summary Questions

1. What does 'diffusion' mean?
2. What is a capillary?
3. How is oxygen carried in the blood?
4. Describe how the structure of the lungs makes them adapted for gaseous exchange.
5. How do the cells in the lining of your lungs work together to keep them clean? Why do smokers develop a cough?

KEY WORDS

trachea
larynx
bronchi
bronchiole
alveolus
gaseous exchange
diffuses
emphysema
mucus
cilia

Inhaled and Exhaled Air

B1.7

- How is inhaled air different from exhaled air?
- How can we compare inhaled and exhaled air?
- How do we get energy from our food?

The air we breathe in (**inhaled** air) is different from the air we breathe out (**exhaled** air).

You will often hear people say that we breathe out carbon dioxide, but is this strictly true?

Comparing inhaled and exhaled air

activity

Comparing oxygen

- Light a small candle and put a gas jar over it.
- Time how long the candle burns.
- Repeat this four more times and work out the average time.
- Now repeat the experiment five more times but using exhaled air. The diagram shows how to do this.
- Make a table of your results.
- What gas is used by the candle as it burns?
- What do the results tell us about how inhaled air is different from exhaled air?
- Why did you repeat each test five times?

How to collect a gas jar of exhaled air

Help Yourself

If you are asked about differences between inhaled and exhaled air always use words that **compare** *them. For example, say 'Exhaled air is warmer', rather than 'Exhaled air is warm'.*

The table summarises the differences between inhaled and exhaled air:

	Inhaled air	Exhaled air
Oxygen	21%	16%
Carbon dioxide	0.04%	5%
Nitrogen and other gases	79%	79%
Water vapour	varies	saturated
Temperature	room temperature	35°C

activity

Comparing carbon dioxide

- Set up the apparatus as shown in the diagram.
- **Gently** breathe in and out of the mouthpiece.
- Observe what happens to the limewater in each boiling tube.
- On a diagram, show how air goes in and out of the apparatus.
- Describe what happens to the limewater in each boiling tube.
- What do the results tell us about how inhaled air is different from exhaled air?

Safety: Wear eye protection. Take care not to suck limewater into your mouth.

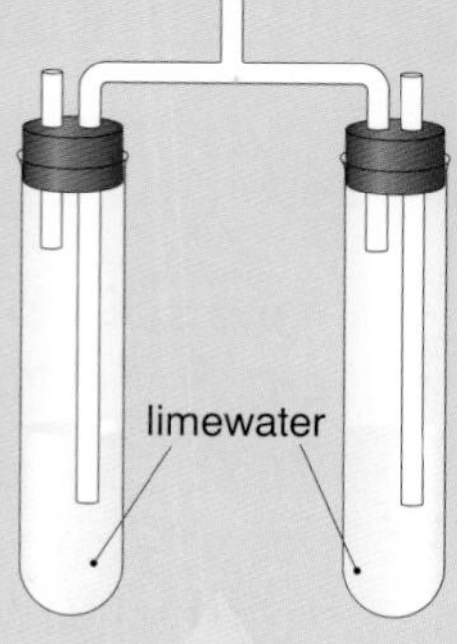

Comparing carbon dioxide

activity

Comparing water vapour

Cobalt chloride paper is normally blue. It goes pink when it gets damp.

- Hold a piece of cobalt chloride paper in the air, well away from your face.
- Breathe out a few times onto another piece of cobalt chloride paper.
- Compare the two pieces of paper.
- What do you observe?
- What do the results tell us about how inhaled air is different from exhaled air?

Safety: Avoid a lot of skin contact with cobalt chloride paper. Wash your hands afterwards.

Why is oxygen so important?

Oxygen is used up when we release energy from food. This happens in a process called **aerobic respiration**:

$$\text{glucose} + \text{oxygen} \longrightarrow \text{carbon dioxide} + \text{water}$$

Summary Questions

1. Copy and complete:

   Inhaled air contains nitrogen, …, carbon … and … vapour.
   Exhaled air contains less … and more … … and ….
   Exhaled air is also … than inhaled air.

2. The chemical formula of glucose is $C_6H_{12}O_6$. What is the formula of:
   a) oxygen, b) water, c) carbon dioxide?
   d) Write a balanced symbol equation for aerobic respiration.

KEY WORDS

inhale
exhale
aerobic respiration

The Heart and Circulation

B1.8

- How is blood carried around the body?
- What types of blood vessels are found around the body and what do they do?

We need oxygen to release energy from glucose. Your blood carries oxygen from your lungs to all your cells. The carbon dioxide made in your cells is carried to your lungs so you can breathe it out.

Your **heart** is made of muscle which contracts, pumping blood around the body.

activity

Taking your pulse

Find your pulse by gently placing your index and second finger on your wrist behind your thumb. Your pulse shows how fast blood is being pumped around your body.

Help Yourself

When you look at a diagram of your heart you will see that the right side seems to be on the left and vice versa! Turn the page round and hold it to your chest. Now you should see that it is correct!

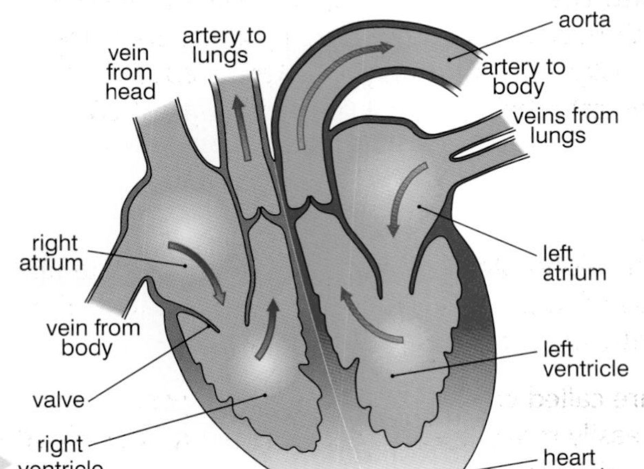

The heart

Your heart is actually two pumps. The right side of your heart collects blood that has passed around your body. Most of the oxygen has been removed from the blood so it is called **deoxygenated** blood.

18 MILES
I wish my legs had cardiac muscle!
211
315
27

The right side of your heart pumps blood to your lungs. It gets rid of the carbon dioxide it has picked up from the cells of the body. It picks up oxygen in the lungs and becomes **oxygenated** blood. This is then carried back to your heart. The left side of the heart then pumps blood to all parts of your body.

Our heart pumps our blood twice, once to the lungs then again around the body. So we call it a double circulation.

The heart has valves. These make sure that blood always flows the right way.

activity

Beating muscle

Hold your hand out. Clench and unclench your fist once per second. Keep on doing this for five minutes. Does your hand get tired?

Luckily your heart is made of special muscle, called **cardiac muscle**, which does not get tired.

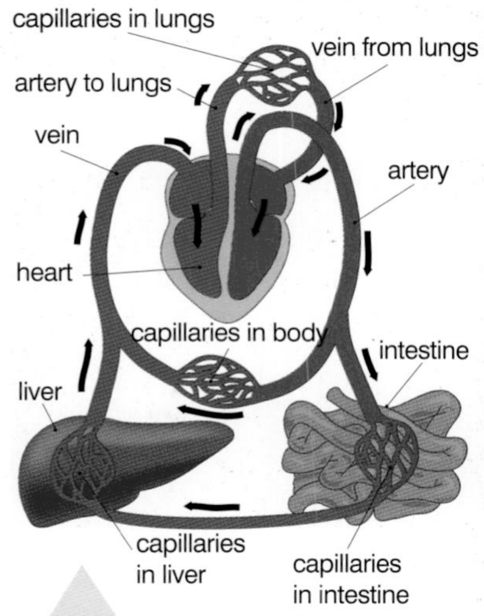

Double circulation

a The diagram on the opposite page shows that the left side of the heart has thicker muscle than the right side. Can you explain why?

b What is the job of the valves in your heart?

Sometimes a person's heart stops working properly. They may have a heart transplant. The first heart transplant was carried out in 1967 by Professor Christiaan Barnard in South Africa. Sadly the patient, Louis Washkansky, died after only 18 days. In 2006 Tony Huesman was still alive after being given a heart transplant 28 years earlier.

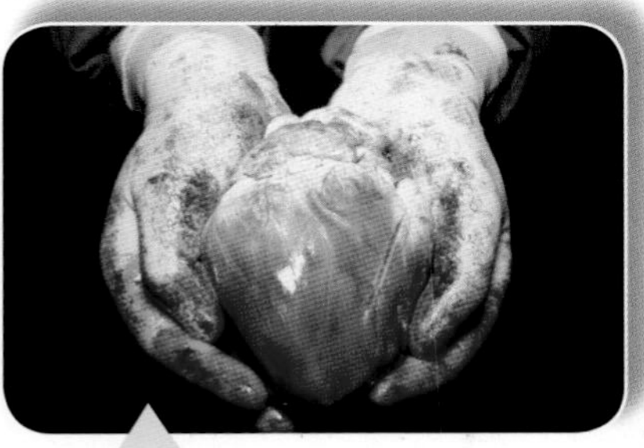

A heart, ready to be transplanted; this gives an idea of the size of your heart

Blood vessels

Blood leaves your heart along blood vessels called **arteries**. Arteries have a thick wall made of muscle. This helps to pump blood as the artery branches out into smaller and smaller vessels.

The smallest vessels are called **capillaries**. These have a very thin wall. Substances can easily move in and out of the blood, but blood cells stay within the capillaries.

Capillaries join together to make larger vessels called **veins** which carry blood back to the heart. Veins do not pump blood so they have a thinner wall.

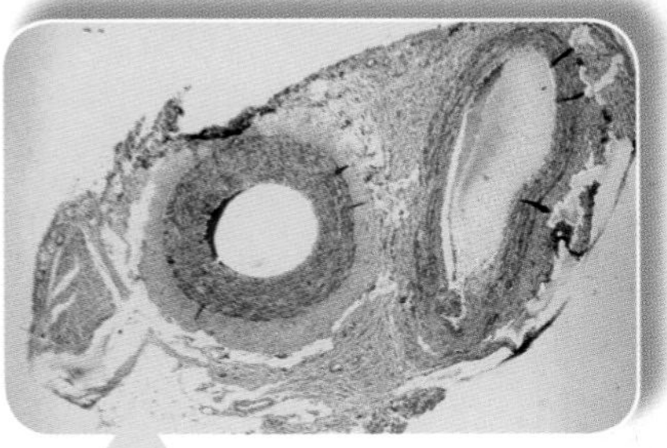

An artery (left) and a vein (right)

Summary Questions

1. Your heart beats an average of 72 times per minute. How many times will it beat if you live to be 85 years old?
2. Most arteries carry oxygenated blood while most veins carry deoxygenated blood. Which arteries and veins are exceptions to this rule?
3. Explain the functions of arteries, veins and capillaries.
4. On a diagram, show the route of the blood through the heart.

KEY WORDS

heart
deoxygenated
oxygenated
cardiac muscle
arteries
capillaries
veins

Supplying the Cells

- How have our ideas about circulation developed over time?
- What substances are carried in the blood?
- How are different substances carried?

Capillaries

Blood takes useful things to cells and removes waste. Capillaries are the smallest blood vessels. They have very thin walls so that oxygen, glucose and carbon dioxide can easily pass through them. The network of capillaries is so vast that every cell in your body has one or more capillaries very close to it.

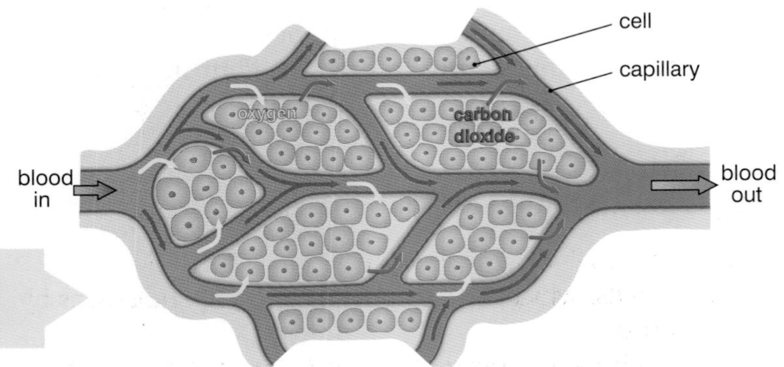

All cells have capillaries nearby

Medicine through time

Galen was a Greek doctor who lived from 129 to 200 AD. He dissected animals because it was illegal to dissect human bodies. He observed arteries and veins but could not see capillaries. He believed that blood passed from one side of the heart to the other through tiny holes in the wall. He thought that blood flowed back and forth along arteries and veins. Galen's writings were used as the basis for European medicine for over 1000 years.

Andreas Vesalius (1514–1564) was a Belgian doctor who dissected human corpses. However, when he showed that there were no holes between the two sides of the heart, most people refused to believe that Galen was wrong!

William Harvey (1578–1657) was an English doctor who showed that blood only flowed one way in arteries and veins. He showed that valves in veins stopped the back flow of blood. He worked out that blood circulated around the body. Even though he could not see capillaries he knew there must be some way for blood to pass from arteries to veins. In 1924 a book was discovered in Egypt, written in 1242 by Ibn al-Nafis, an Arab doctor. It showed that he knew blood circulated through the lung over 300 years before William Harvey's discoveries.

William Harvey's drawing showing valves in the veins

Blood cells

Oxygen is carried by red blood cells. These contain a chemical called **haemoglobin**. In your lungs, oxygen binds to haemoglobin to make oxyhaemoglobin. When red blood cells reach body tissues, oxygen is released from oxyhaemoglobin. It passes through the capillary walls and into cells that need it.

a Where is haemoglobin found?

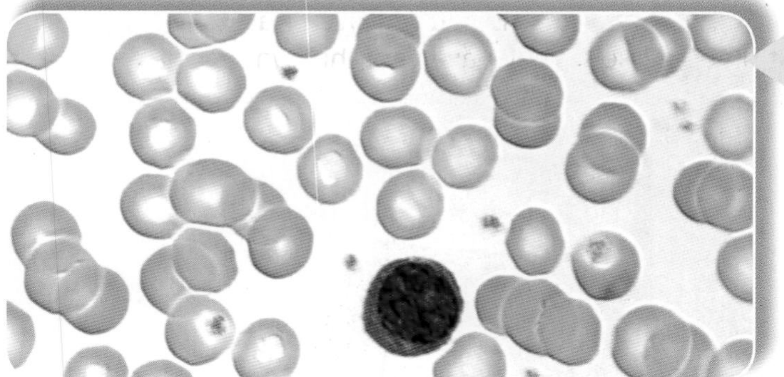

Blood cells

The photograph shows what blood is made of:

- **red blood cells,** which carry oxygen around the body
- **white blood cells,** which help to protect us against diseases by killing microbes
- **platelets,** which help to make the blood clot so that we stop bleeding
- **plasma**, which is the liquid part of the blood. It carries digested food such as glucose, minerals, and amino acids. When blood reaches the capillaries some of the plasma leaks out and forms **tissue fluid**. Nutrients diffuse into the cells as tissue fluid surrounds them. Plasma also carries carbon dioxide and other waste such as urea. Urea is removed from the blood in the kidneys.

Summary Questions

1. Name two substance that pass from blood to cells and one substance that passes from cells to blood.
2. What is tissue fluid?
3. Olympic athletes can run 100 m in less than 10 s. Why can't athletes run at this speed for 1500 m?
4. Copy and complete the table to summarise the functions of the different parts of the blood:

Part of the blood	Function
Red blood cells	
White blood cells	
Platelets	
Plasma	

Help Yourself

Can you remember how red blood cells are adapted for carrying oxygen? Their shape gives them a large surface area. They have no nucleus so there is more space for carrying haemoglobin.

science@work

Scientists often use microscopes to examine blood cells in the diagnosis of diseases such as sickle cell anaemia and leukaemia.

KEY WORDS

haemoglobin
red blood cells
white blood cells
platelets
plasma
tissue fluid

Exercise

B1.10

- What happens to our heart and breathing rate when we exercise?
- What happens if we cannot get enough oxygen to our muscles?

Imagine you have just been out jogging. How do you feel afterwards? How does exercise affect your body?

a Write down some changes you observe in your body after you have been exercising.

Exercising releases a lot of energy

activity

The effects of exercise

- Sit quietly for a few minutes then take your pulse. This is called your resting pulse. The diagram should help you to find your pulse. To make it easier you can count your pulse for 15 seconds then multiply this number by four, to find how many times your heart beats in one minute.
- Exercise for one minute.
- Take your pulse again every minute until it has reduced to your resting pulse.
- Repeat with another person.
- Put your results in a table. Draw a graph so that you can compare how both people were affected by exercise.
- How will you make sure your investigation is as fair as possible?
- What did your results show?
- Who do you think was fitter?
- Explain how you could improve your investigation.

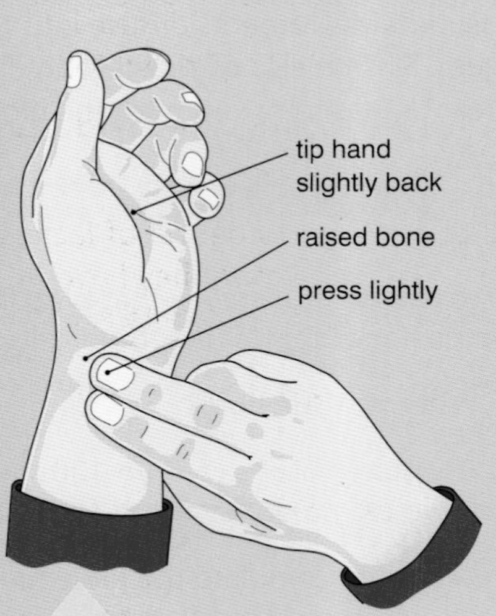

Your pulse is found where an artery passes close to your skin

When you exercise, your muscles need more energy. Energy is released when glucose reacts with oxygen. So the more you exercise the more glucose and oxygen you need in your muscles. You breathe faster and deeper so more oxygen can get into your blood. Your heart beats faster so more blood is carried to your muscles.

Releasing more energy also produces more carbon dioxide. This is removed by being carried by blood to the lungs and breathed out.

'Aerobics' is any type of exercise which increases your heart rate but is slow enough to let you get enough oxygen around your body. Exercises that you can keep doing for a long time such as walking, jogging and cycling are aerobic. Some exercise, such as sprinting, uses so much oxygen that your lungs and blood cannot get enough to your cells.

b Give some examples of aerobic exercise.

Oxygen debt

Anaerobic respiration

If you do very vigorous exercise your heart beat and breathing will reach their maximum rate. Your muscle cells will not be able to get any more oxygen. When this happens your cells can release energy for a short while by **anaerobic respiration**. Anaerobic means 'without air':

$$\text{glucose} \longrightarrow \text{lactic acid} + \text{carbon dioxide}$$

You cannot respire anaerobically for long because:

- much less energy is released than in aerobic respiration,
- lactic acid is **toxic**.

Your body needs to get rid of lactic acid. Athletes continue to gasp for breath, even when they have finished their race. They are in **oxygen debt** and need to take in oxygen to break down lactic acid.

Summary Questions

1. Explain why your heart and breathing rate increase when you exercise.
2. Explain how to take your pulse. A drawing would help.
3. Draw a table to summarise the differences between aerobic and anaerobic respiration.
4. Why isn't it possible to respire anaerobically for long periods of time?

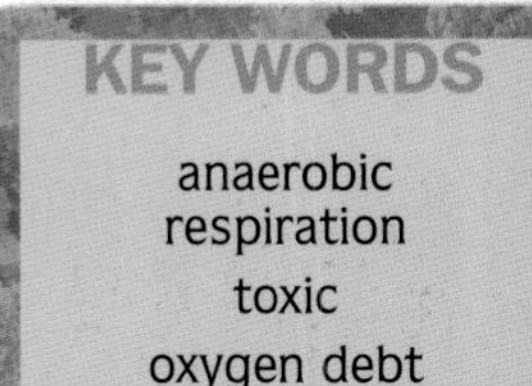

KEY WORDS

anaerobic respiration

toxic

oxygen debt

Excretion and Homeostasis

- How do we control our temperature?
- How do we control the amount of water in our body?
- How do we remove toxins from our body?

Homeostasis

A car engine works best at a certain temperature. When it first starts on a cold morning it doesn't run perfectly, but after a while it warms up. Once it is at the right temperature, water in the radiator stops it getting any warmer. A leaky radiator causes an overheated engine and big problems.

Your body needs the right conditions to work properly. Control of temperature, the amount of water and sugar and the pH of your body is called **homeostasis**.

A car engine has a thermostat to control its temperature. Your brain contains cells that measure the temperature of your blood. Your body temperature should be about 37°C. If it gets too warm these cells 'switch on' cooling systems. If it gets too cold they 'switch on' warming systems.

a Where is your body's 'thermostat'?

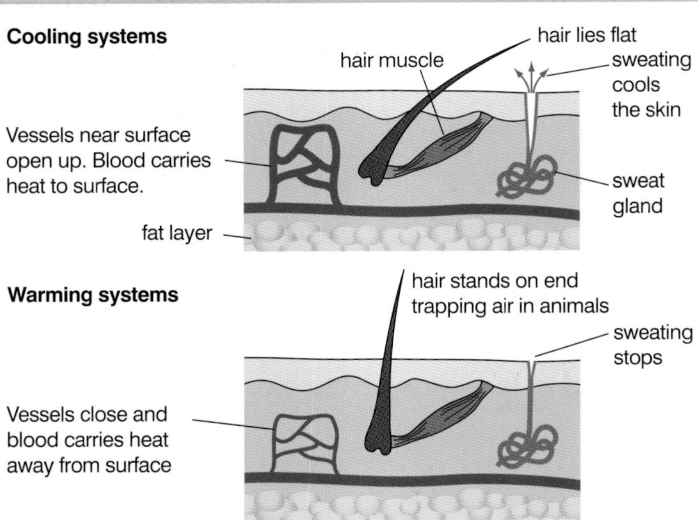

Cooling systems and warming systems

activity

Does sweating actually cool us down?

Design an investigation to find out if sweating actually cools us down. Use a boiling tube to model a body, with paper towels as skin. Use hot water in the boiling tube to model blood.

- What will you have to do to make it fair test?
- How will you make sure your results are as reliable as possible?

When you have done your investigation report back to the rest of your class what you have found out.

All humans have a layer of fat under their skin. This is used as a store of energy but it also insulates us.

If you are waiting for a bus on a cold day you might start shivering. This rapid muscle movement produces heat and warms you up!

Controlling body water

On a hot day you produce a lot of sweat, especially if you are exercising. You need to drink plenty of water to replace what you have lost. In addition your kidneys control the water in your body:

- If you have too little water in your body you make less urine. Your urine is a darker colour because it is more concentrated.
- If you have too much water you make more urine. It is a paler colour as it is less concentrated.

Did You Know?

Sweat doesn't smell. Bacteria living on your skin make the smell when they feed on chemicals in your sweat. Sweat contains most of the chemicals found in urine!

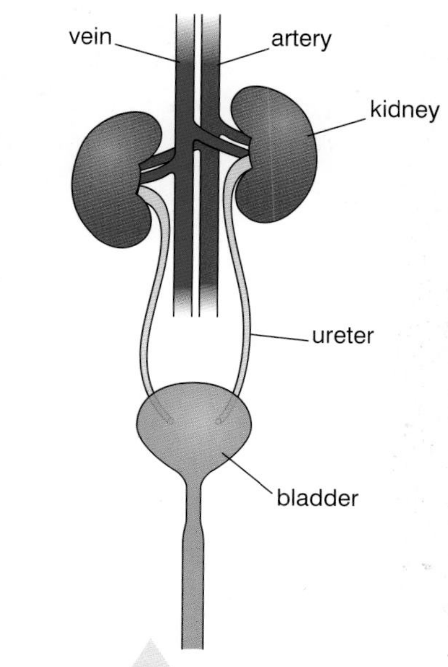

Kidneys and bladder

Excretion

Your kidneys also remove chemical waste from your body. A toxic substance called **urea** is made in your cells. This is removed from your blood by your kidneys as urine. Urine is carried to your bladder and stored until you are ready to go to the toilet. Removing toxic substances made in cells is called **excretion**.

b When would you expect to make the most urine, in the summer or winter?

c Which is a darker colour, summer or winter urine?

Summary Questions

1. What does 'homeostasis' mean?
2. Why does your skin look red when you are hot?
3. Copy and complete the table to show how your body keeps you warm or cool:

	Keeping warm	Keeping cool
Hairs		
Sweat glands		
Surface blood capillaries		
Muscles of body		

4. Copy and complete:

   If we sweat a lot we make less … to keep our body water balanced. Urine is made in our … . It contains a toxic chemical called … . Urine is stored in our … until we go to the toilet.

KEY WORDS

homeostasis
urea
excretion

The Nervous System

- What are reflexes and how do they help to protect us?
- What are our senses?
- What stimuli are detected by our skin?

Reflexes

Have you ever touched something which you didn't expect to be hot? You probably moved your hand away pretty quickly, without even having to think about it. Also it probably didn't start to hurt until after you moved your hand away. This is an example of a **reflex action** (or just a reflex).

Reflexes help to protect us

A reflex arc

Reflexes help to protect us from injury. They happen very quickly and they are automatic – you don't have to think about them.

In a reflex, an electrical impulse passes very quickly along the nerves. A chemical carries the message between nerves. The fewer nerves it has to pass along, the faster the message is carried. The diagram shows a reflex arc.

You only start to feel pain when a message reaches your brain. That doesn't happen until after you have moved your hand.

activity

Measuring reaction time

You can use a ruler to measure your reaction time:

- Hold your hand horizontally in front of you.
- Your partner holds a metre ruler upright with the zero level with your hand.
- Your partner lets go of the ruler (without telling you).
- Catch the ruler as quickly as you can.
- Read off the measurement, level with your hand.
- Write this in a table. Repeat nine more times.
- Swap over so that you drop the ruler and your partner catches it.
- Did your reaction time improve with practice?
- Who had faster reactions?
- Find out which hand, right or left, gives you the fastest reaction time.

a Moving your hand away when you touch something hot is a reflex action. Think of three more examples.

b Why is it important that reflexes are fast and automatic?

Reaction time is important in many situations

Senses

A reflex happens when your body responds to an external **stimulus** which is harmful. These are detected by your sense organs.

We have five sense organs:

- **skin** – detects touch, pressure, heat, cold and pain
- **nose** – detects chemicals in the air ⟶ sense of smell
- **tongue** – detects chemicals in food ⟶ sense of taste
- **eyes** – detect light ⟶ sense of sight
- **ears** – detect vibrations in the air and movement ⟶ sense of hearing and balance.

Did You Know?

An earthworm's whole body is covered with taste receptors. A star-nosed mole has more touch receptors in its nose than you have in your hand!

activity

Sensitive skin

Your skin contains touch receptors. Do all parts of the skin have the same number of touch receptors?

- Bend a hair pin so the tops are 5 mm apart.
- Blindfold a partner.
- Touch your partner on the finger tips with one or two tips of the hair pin.
- Your partner has to say whether they can feel one or two tips.
- Repeat nine more times and record the number of times your partner is correct.
- Repeat on other parts of the hand and arm.
- What does the investigation tell us about the number of touch receptors in different areas of the skin?

Skin is sensitive to different stimuli

Summary Questions

1. How does pain protect us, even when the danger is over?
2. Give two reasons why a reflex is fastest when it only passes along a few nerves.
3. Draw a concept map to summarise the functions of the sense organs.

KEY WORDS

reflex action
stimulus

know your stuff

▼ Question 1 (level 3)

Look at the diagram of the digestive system.

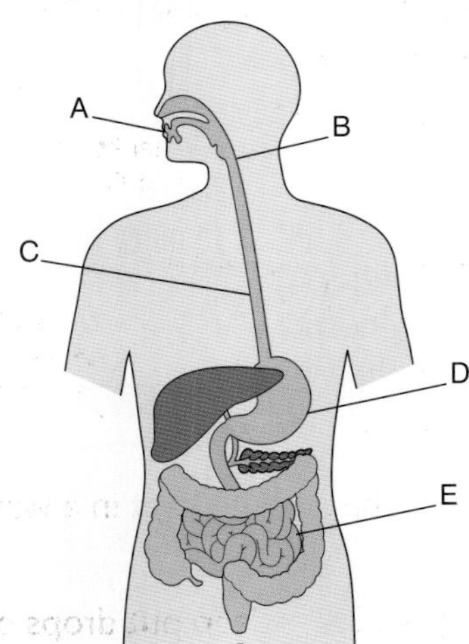

a (i) Which letter labels the stomach? [1]

(ii) Which letter labels the small intestine? [1]

(iii) Which letter labels the oesophagus? [1]

(iv) Glucose is absorbed in the small intestine.

How is glucose carried from the intestine to other parts of the body? [1]

b Some footballers take glucose tablets before a match. Why? [1]

c The table below shows what four people ate for lunch:

Name	Lunch
Runi	chicken salad
Faye	double cheeseburger and chips
Sitel	cola drink and a biscuit
Barbara	lentil soup and an orange

(i) Who had the most sugar in her lunch? [1]

(ii) Who has the most fat in her lunch? [1]

(iii) Barbara says eating too much fat is bad for you.

Give *one* reason why fat is bad for you. [1]

▼ Question 2 (level 4)

a Lynn measures her pulse rate and then swims 20 lengths of the pool. Afterwards she measures her pulse again. How would the swim affect her pulse rate? [1]

b The list shows three useful substances and two waste products. They are all in the blood.

oxygen carbon dioxide glucose
vitamins urea

Which *two* of these are waste products produced by the body? [2]

▼ Question 3 (level 5)

The following are important parts of a balanced diet:

carbohydrates proteins fats
vitamins water

a Use the words above to copy and complete these sentences:

(i) An orange is a good source of … and …. [2]

(ii) An orange is a poor source of … and …. [2]

b Name *two* parts of a balanced diet which are *not* in the list above. [2]

c In order to obtain the nutrients, food must be chewed properly. Give *two* reasons why proper chewing is important. [2]

How Science Works

▼ Question 1 (level 6)

The table shows the amount of six different nutrients in three types of mashed potato.

Kidzmash is an instant mash, specially made for children.

Nutrient	100g Mashed potatoes	100g Instant mash made up ready to eat	100g 'Kidzmash' made up ready to eat
Carbohydrate (g)	12.0	12.6	12.5
Protein (g)	1.4	1.4	1.4
Fat (g)	0.1	0.1	0.1
Vitamin C (mg)	20	10	60
Fibre (g)	1.3	1.2	1.2
Sodium chloride (mg)	75	80	78

a A scientist compared the three types of mash. Why was it a fair comparison? [1]

b All types of mash include some sugar. Why isn't it listed in the table? [1]

c What ingredient helps the movement of food through the intestines? [1]

d What evidence is there that vitamin C is added when making Kidzmash? [1]

e A boy said, "There is more sodium chloride than protein in mashed potato". How can you tell from the table that the pupil is wrong? [1]

▼ Question 2 (level 7)

Jasdeep and Mandeep carried out an investigation into the digestion of starch by the enzyme called amylase.

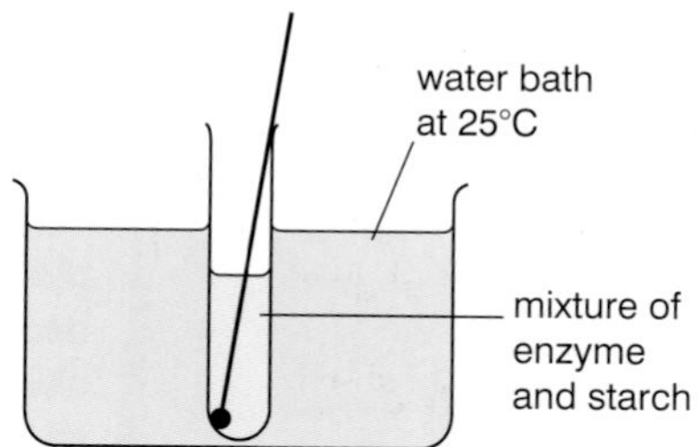

a Why was the mixture kept in a water bath? [1]

Jasdeep and Mandeep put drops of iodine solution on a white tile.

Every 30 seconds they added a drop of the mixture of enzyme and starch to a drop of iodine solution. At first the drops turned blue, but after four minutes they stayed brown.

b Why did the mixture stop turning the drops of iodine solution blue after four minutes? [1]

c They then carried out the experiment with the water bath at 35°C. This time, the drops stopped turning blue after two minutes. Explain why. [1]

d Sadie and Tom want to compare the experiment at 35°C with the results from the experiment at 25°C. Describe what they need to do to make this a fair test. [1]

B2.1

Ecology

The importance of plants

Charlie wants to brighten up his lounge. He thinks some plants would make it look a lot nicer. He talks to his friend, Freddie.

Freddie and Charlie do not seem to know very much about growing plants!

a Write a letter or e-mail to Charlie giving him some advice about what plants need in order to grow well.

Plants are essential for life on Earth. Animals, including humans, need plants for food and to produce the oxygen they breathe in.

b Write a chemical equation for the process of respiration.

All our energy comes from eating plants, or from eating animals which eat plants.

c Where do plants get their energy?

d Draw a simple plant and label these parts: leaf, stem, flower, root. Write a sentence to describe the function of each part you have labelled.

e Draw a diagram of a plant cell and label these parts: cell wall, cell membrane, nucleus, cytoplasm, vacuole, chloroplast. Write a sentence to describe the function of each part you have labelled.

Freddie and Charlie decide to look at the plants growing in the garden.

f How do you think the berry bush on the right of the picture started to grow?

Animals and plants depend on each other. We say they are interdependent. Animals could not survive without plants. Plants' survival is sometimes helped by animals.

activity

Studying a garden

- Look carefully at the picture of the garden above. It shows lots of animals and plants.
- Write down how the plants can help the animals to survive.
- How might the animals help the plants?

Making Food

B2.2

- How can we test a leaf for starch?
- How do plants make glucose and other substances that contain carbon?
- How can we show the presence of starch in the leaves of plants?

activity

Testing leaves for starch

We can use iodine to test for starch.

- Take a leaf from a plant that has been in a sunny place.
- Put the leaf in a beaker of boiling water from a kettle. Leave it for about two minutes. This breaks down the cell membrane so that iodine can get into the cells.
- Put the leaf in a test tube of ethanol. Stand the test tube in the beaker of hot water. This removes the colour from the leaf so that we can see the colour of the iodine clearly.
- When you have removed the colour from the leaf, take it from the ethanol. Rinse the leaf in cold water.
- Spread out the leaf on a white tile.
- Cover the leaf with iodine solution.
- A dark blue/black colour shows that the leaf contains starch.
- Why was the leaf put in hot water?

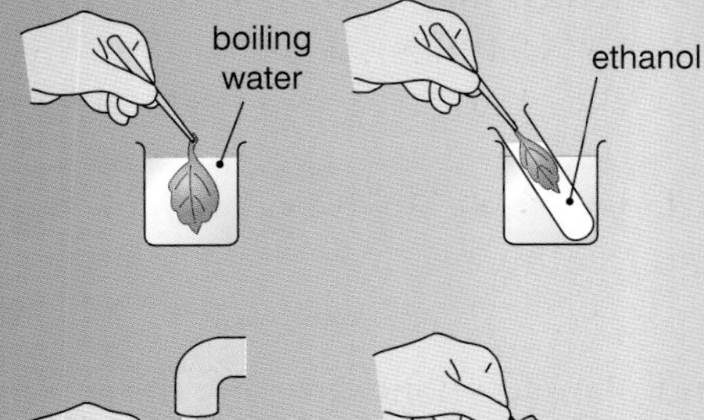

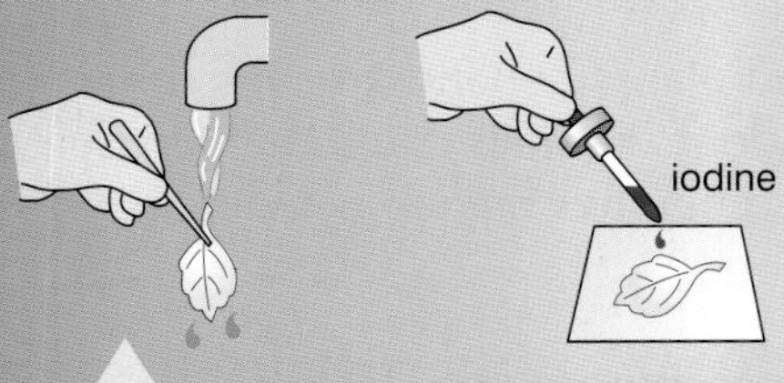

How to test a leaf for starch

- Why did you remove the colour from the leaf?

Safety: Wear eye protection. No naked flames – ethanol is highly flammable.

You should already know that plants need light to grow.

activity

Is light needed to make starch?

- Take a plant that has been kept in a dark place for about three days. Keeping a plant in a dark place makes sure its leaves do not contain any starch.
- Cover part of a leaf with aluminium foil or black paper.
- Leave the plant in a sunny place for at least a day.
- Test the plant for starch as before.

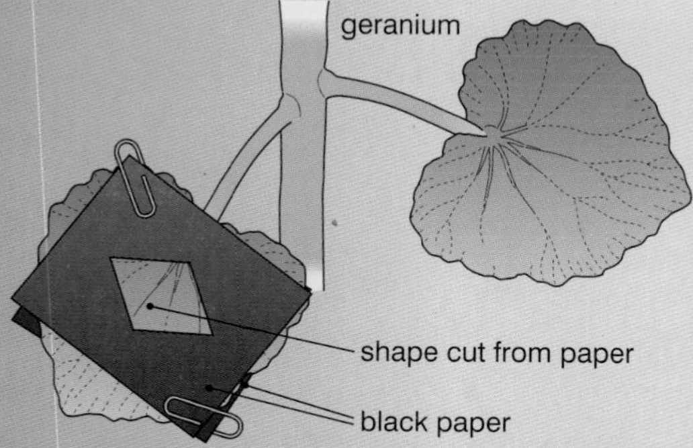

- Make a drawing of the leaf after you have added iodine. Compare this with the shape of the cover you put over the leaf.
- What does this experiment tell us?

Safety: Wear eye protection. No naked flames – ethanol is highly flammable.

Plants make glucose in a process called **photosynthesis**. 'Photo' means 'with light'. 'Synthesis' means 'to make'. Plants convert some of the glucose into starch, which is stored in the leaves. Some glucose is carried to other parts of the plant where it can be stored as starch, e.g. a potato some glucose is made into other substances.

a What does the word 'photosynthesis' mean?

Help Yourself

Look at the work you did on digestion in B1. Can you remember how we digest starch by breaking it down into glucose molecules? Plants make starch by joining together lots of glucose molecules.

Jan Baptista van Helmont was a scientist who lived from 1580 to 1644. He did an experiment to investigate the growth of a plant. He put 200 lb of dry soil in a pot. He planted a young tree weighing 5 lb. He kept the tree watered but added nothing else. After five years he weighed the plant. It weighed 169 lb. He weighed the soil and found it had lost only a tiny amount of weight. He concluded that the tree had grown from the water he gave the tree!

Summary Questions

1. We do not have time to do an experiment that takes five years! Can you think of a way of using a fast-growing plant such as cress to carry out a similar investigation to Van Helmont.
2. Van Helmont concluded that the tree grew from the water he added to the pot. Do you think he was right? Write a letter to Van Helmont explaining how you think the tree grew.

KEY WORDS

photosynthesis

Photosynthesis

- Do plants need chlorophyll for photosynthesis?
- What are the raw materials needed for photosynthesis?
- Apart from glucose, what else is made in photosynthesis?

Variegated leaves

Being green

The green colour of a leaf is due to a substance called **chlorophyll**. In some plants only part of the leaf has chlorophyll. These leaves are called variegated leaves.

activity

Do plants need chlorophyll for photosynthesis?

- Take a variegated plant that has been kept in the dark for two days.
- Leave it in a sunny place for a few hours.
- Take a leaf from the plant. Draw the leaf showing which parts are white and which are green.
- Test the leaf for starch as on page 30.
- Draw a diagram of the leaf after you added iodine. Compare it with the drawing you did before.
- What does this experiment tell you about the importance of chlorophyll in photosynthesis?

Safety: Wear eye protection. No naked flames – ethanol is highly flammable.

activity

Do plants need carbon dioxide for photosynthesis?

Soda lime is a chemical that absorbs carbon dioxide.

- Take a plant that has been kept in a dark place for two days.
- Put a small dish of soda lime on top of the soil.
- Put a plastic bag over the whole plant and use an elastic band to hold it in place.
- Leave it in a sunny place for a few hours.
- Test one leaf for starch.
- What control should you use in this experiment?
- What conclusion can you make from this experiment?

Safety: Soda lime is corrosive so wear chemical splash-proof eye protection. No naked flames – ethanol is highly flammable.

a What is meant by a 'variegated' leaf?

b Why did you use soda lime in one of the above experiments?

c Name three things that are needed for photosynthesis to happen.

activity

Making oxygen in photosynthesis

- Put some *Elodea* (Canadian pondweed) in a large beaker.
- Cover the *Elodea* with a glass funnel.
- Put a test tube of water over the funnel.
- Put the beaker in bright light.
- After a few minutes you will see bubbles of gas coming from the *Elodea*.
- After a few hours there should be enough gas to test.

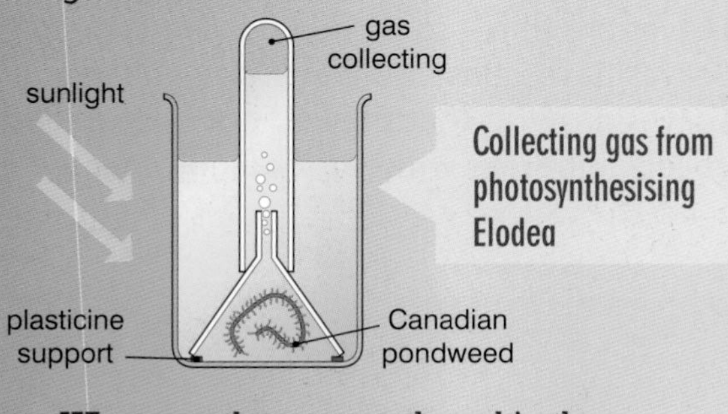

Collecting gas from photosynthesising Elodea

- What was the gas produced in the experiment?

Stretch Yourself

Plants also use water in photosynthesis. We cannot easily show this by doing an experiment. If a plant does not get water it will die anyway as all of the processes in its cells need water. Scientists can show that water is used to make starch. They use special water molecules which are labelled with heavy oxygen. They can then detect these atoms by a special test. If the plant is given this 'labelled water', they can show that it is used in photosynthesis. The hydrogen atoms are found in the starch that is made and the oxygen is given out as oxygen gas.

d What test is used to detect oxygen?

Summary Questions

1. Gardeners often use a paraffin heater to keep their greenhouse warm. Paraffin is a hydrocarbon (it is made from hydrogen and carbon atoms). Give *two* reasons why using the paraffin heater will help the plants to grow.
2. Make a poster to summarise the experiments you have done to investigate photosynthesis.
3. Copy the table to summarise the process of photosynthesis. Put the words below in the correct column:

glucose oxygen light chlorophyll carbon dioxide water

Raw materials (reactants)	Energy source	Energy collector	Products

KEY WORDS

chlorophyll

Leaves and Photosynthesis

- How are leaves adapted for photosynthesis?
- What factors affect the rate of photosynthesis?
- How can I plan a fair test?

Rainforest

It is dark under the leaves of the rainforest

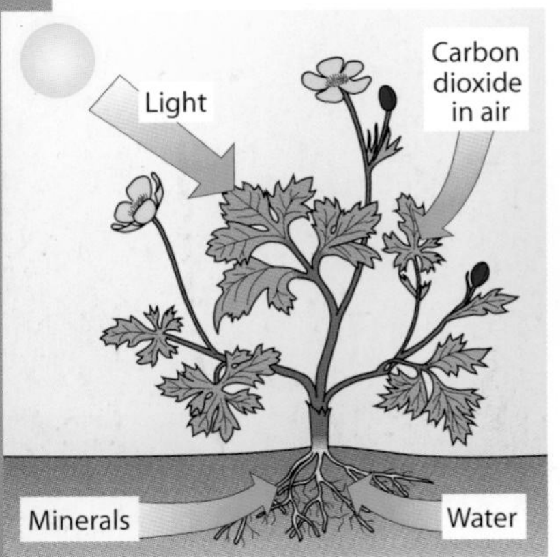

Absorbing light

The trees in the photographs have grown so that all we can see from the air are leaves. This makes sure that the leaves absorb as much light as possible.

Photosynthesis occurs in leaves:

$$\text{carbon dioxide} + \text{water} \xrightarrow[\text{chlorophyll}]{\text{light energy}} \text{glucose} + \text{oxygen}$$

$$6CO_2 + 6H_2O \longrightarrow C_6H_{12}O_6 + 6O_2$$

Leaves are adapted so they photosynthesise efficiently:

- Leaves are broad and flat and grow at 90° to the Sun's rays – so they can collect as much light energy as possible.
- Leaves are thin – so carbon dioxide can get to the cells easily.
- Leaves have veins – so water gets to the leaf cells.
- Leaves grow so that they do not overlap each other – so they all collect light rays without shading other leaves.

activity

Microscopic leaf structure

- Use a microscope to examine a section of a leaf.
- Draw what you see and try to identify the parts labelled in the diagram below.
- Paint a small part (about the size of a fingernail) of the upper and lower surfaces of a leaf with clear nail varnish.
- When it is dry, peel off the nail varnish. The nail varnish contains an imprint of the leaf surfaces.
- Look at the nail varnish through your microscope. Compare the two surfaces. Can you see the stomata?
- Draw what you see.

Safety: Use in a well ventilated room. Avoid breathing in fumes from nail polish. Nail polish may well be flammable.

a Which gas passes into the leaf cells and is used in photosynthesis?

b Which gas is produced in photosynthesis and passes out of the cells?

activity

Leaf structure

- Examine a leaf with a magnifying glass.
- Draw the leaf.
- **Annotate** your drawing to explain how the leaf is adapted for photosynthesis.

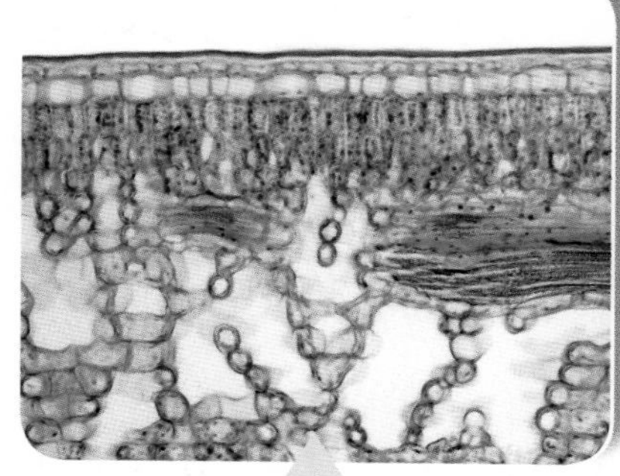

A section through a leaf – can you match the layers with the diagram underneath?

Leaves are made up of several layers:

- Waterproof waxy cuticle – stops the leaf drying out.
- Transparent upper epidermis – light can get to the next layer.
- Palisade layer is made of **palisade cells**, which have lots of **chloroplasts** containing chlorophyll.
- Spongy layer with a lot of air spaces – gases can pass in and out of the cells.
- Lower epidermis has tiny pores called **stomata**, which are opened and closed by **guard cells.**

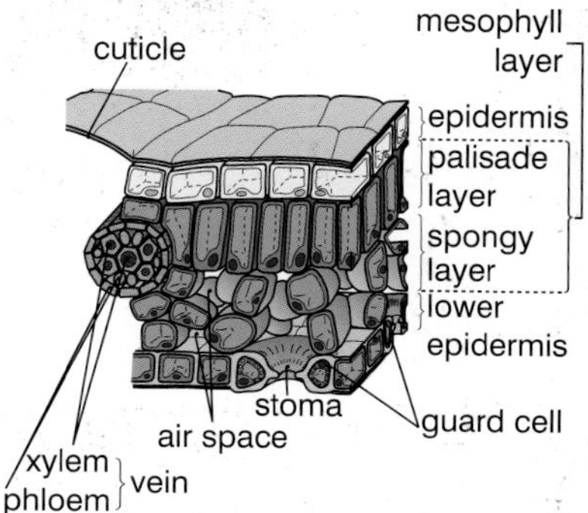

Leaf section

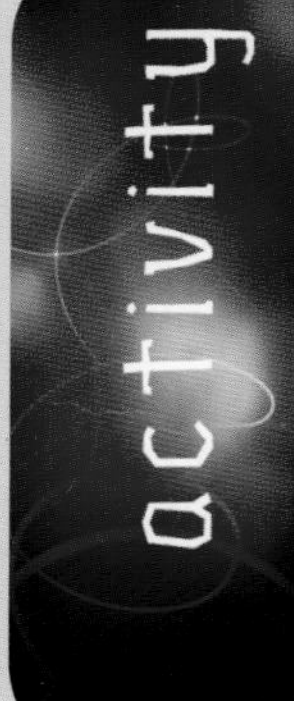

Factors that affect the rate of photosynthesis

Investigate how moving a lamp nearer to an *Elodea* plant affects the rate of photosynthesis. The more bubbles per minute, the faster it is photosynthesising.

- How will you make it a fair test?
- How will you present your results?

Limiting factors

John put *Elodea* in water with different concentrations of carbon dioxide. He counted the number of bubbles per minute. The same piece of *Elodea* and the same light intensity were used each time. Here are his results:

1. Draw a poster or 'mind map' to summarise how the process of photosynthesis takes place in a leaf.

KEY WORDS

palisade cell
chloroplast
stomata
guard cell

Food Chains

- What does a food chain tell us?
- How can we classify the organisms in a food chain?
- What are decomposers?

What happens in a food chain?

Food chains show what eats what in a habitat. They show the flow of energy from one organism to another.

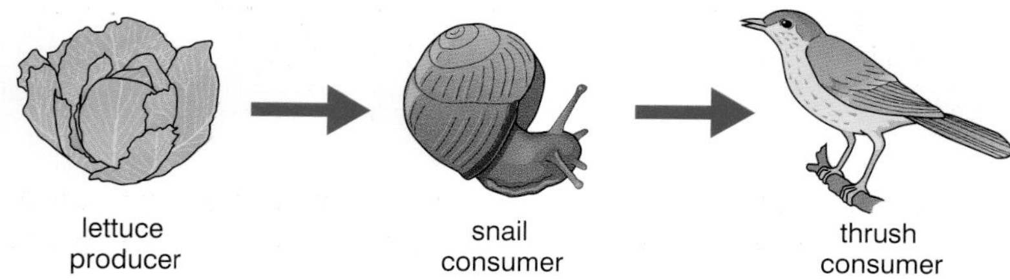

A food chain

This food chain shows that the snails eat lettuce and the thrush eats snails. The arrows in a food chain show the direction of the flow of energy.

a What is the name of the process in which green plants produce food?

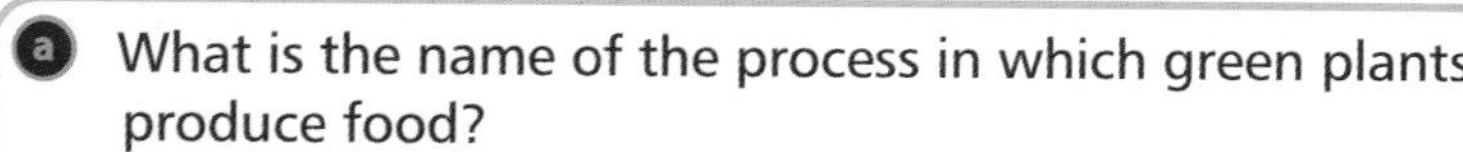

b Write a word equation to show what happens in this process.

A food chain always starts with a green plant because they are **producers**. Producers use light energy to produce food. Snails and thrushes are **consumers**. They eat plants or other animals.

We can classify the organisms in a food chain in other ways:

- **herbivores** – animals that eat plants
- **carnivores** – animals that eat other animals
- **omnivores** – animals that eat animals *and* plants
- **predators** – animals that catch, kill and eat other animals
- **prey** – animals that are caught and eaten by predators.

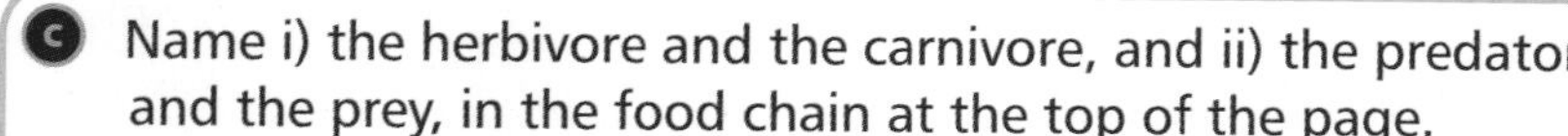

c Name i) the herbivore and the carnivore, and ii) the predator and the prey, in the food chain at the top of the page.

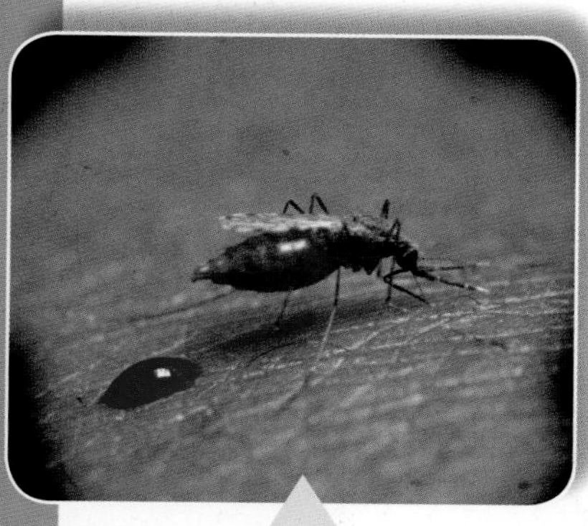

Mosquitoes are parasites

Predators are always carnivores, but not all carnivores are predators. Vultures eat animals but don't catch them – they are **scavengers** that eat animals that are already dead. Mosquitoes suck blood from animals but they don't catch them first. Mosquitoes are **parasites**.

activity

Making food chains

Collect pictures of plants and animals. Make a display of some food chains.

Label the food chains to show which organisms are a) producers or consumers, b) herbivores, carnivores or omnivores, c) predators or prey.

Did You Know?

Vultures have fewer feathers on their heads and necks so they do not get coated in blood as they feed on the internal organs of animals!

Vultures are scavengers

Animals and plants in the food chain die. Their bodies are broken down by **decomposers**. These include bacteria and fungi, as well as animals such as woodlice and earthworms. The bodies of animals and plants contain a lot of **minerals**. When they decompose these minerals are released back into the soil. Decomposers also break down urine and faeces.

d What do 'decomposers' do? Give some examples of decomposers. Where would a decomposer fit into a food chain?

A decomposer

Summary Questions

1. Change these sentences into food chains:
   a) Field mice eat brambles. Weasels eat field mice.
   b) Greenfly eat rose bushes. Blue tits eat greenfly. Sparrowhawks eat blue tits.
2. Why are plants called 'producers'?
3. Why are decomposers so important?

KEY WORDS

food chain
producers
consumer
herbivore
carnivore
omnivore
predator
prey
scavenger
parasite
decomposer
mineral

Food Webs

B2.6

- How do food chains link together to make food webs?
- How do changes in food webs affect populations?

What is a food web?

What did you eat yesterday? You probably had lots of different things otherwise life would be quite boring! Most animals eat a lot of different things too. A food chain does not tell us everything about what an animal eats. Several food chains can be linked together to make a **food web**.

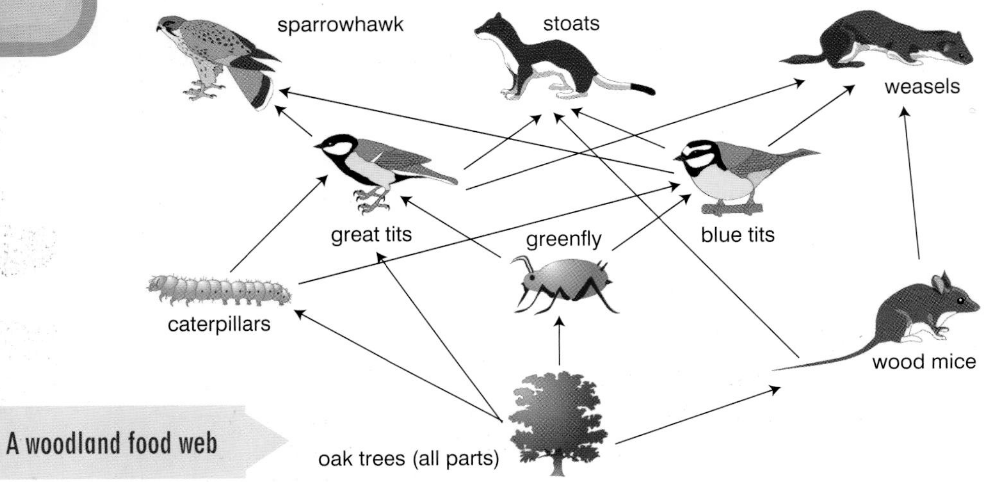

A woodland food web

a Draw five food chains from the woodland food web.

b For each organism, decide if it is a producer or a consumer.

c For each organism, decide if it is a herbivore, a carnivore or an omnivore.

Feeding relationships affect the size of the population of an animal. The population is affected by:

- the amount of food available
- **competition** for food between organisms in a food web. Only a certain population can exist at each level in a food web. For example, stoats and weasels eat the same things so they compete with each other.
- the number of predators.

d Do plants compete with each other? Explain your answer.

Upsetting the web

A change in the size of the population of one species in a food web can have an effect on other species. Look at the woodland food web again.

Imagine a disease killed many of the stoats. The blue tit population would probably increase because fewer of them would be eaten.

Sometimes it is more difficult to predict what might happen. Imagine a disease kills many of the blue tits. Would the great tit population *increase* because there is more food available, owing to fewer being eaten by blue tits? Or would the great tit population *decrease*, because more of them are eaten by predators, which have fewer blue tits to eat?

Look at the woodland food web. A pesticide is used which kills most of the caterpillars.

I have come to mess up your food chains.

e i) What effect would this have on the population of great tits?
ii) Explain your answer.

f i) What effect would this have on the number of greenfly?
ii) Explain your answer.

Alien invaders

No, not green creatures from outer space – an 'alien' invader is any creature which is introduced into a different habitat. Some are introduced accidentally. For example, the Chinese mitten crab reached Europe in the ballast water of ships. This crab competes with native species and could cause a reduction in numbers.

Grey squirrels were introduced to Britain from North America in the 1860s. They compete with native red squirrels, which were once found all over the UK. Now red squirrels only live in a few areas of England and Scotland. Also we find that Sitka spruce trees, introduced into the UK, are not as good a food source for red squirrels.

Red squirrel

Grey squirrel

Summary Questions

1 Read this information about a marine habitat:

Winkles and limpets eat seaweed. Winkles are eaten by octopuses and crabs. Crabs also eat limpets. Starfish eat limpets. Seals eat crabs and octopuses. Seagulls eat crabs and starfish. Killer whales eat seals.

a) Use this information to draw a food web.
b) Name the producer.
c) Name the herbivores.
d) Name the carnivores.
e) If a disease kills most of the crabs, what effect could this have on other members of the food web?

2 Find out about problems caused by alien invaders such as mink, coypu and red signal crayfish.

KEY WORDS
food web
competition

Food Pyramids and Energy Flow

- What are pyramids of numbers?
- How can they show energy flow through ecosystems?
- How does bioaccumulation occur in ecosystems?

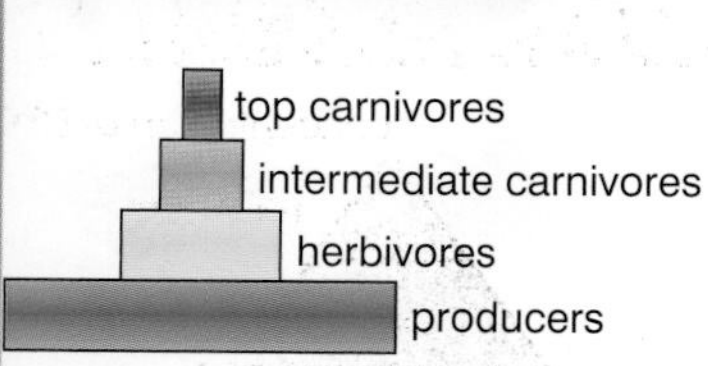

A pyramid of numbers

Pyramids of numbers

Food chains tell us what eats what. **Pyramids of numbers** tell us *how many* organisms there are at each stage.

There are lots of plants at the bottom of the pyramid and fewer herbivores feeding on the plants. Even fewer intermediate carnivores feed on the herbivores and above them, there are even fewer top carnivores, therefore the smallest box of the pyramid.

Food chains show the flow of energy. Pyramids of numbers get narrower at each level as only a small part of the energy is passed on. For example, much of the plant material, like wood, is inedible. Other parts, such as roots, cannot easily be reached by animals.

Much of the energy taken in by an animal is used up to keep warm or for moving around. Some of the energy in food is passed out of the animals in urine and faeces. Only about 10% of the energy taken in is actually stored in an animal's body.

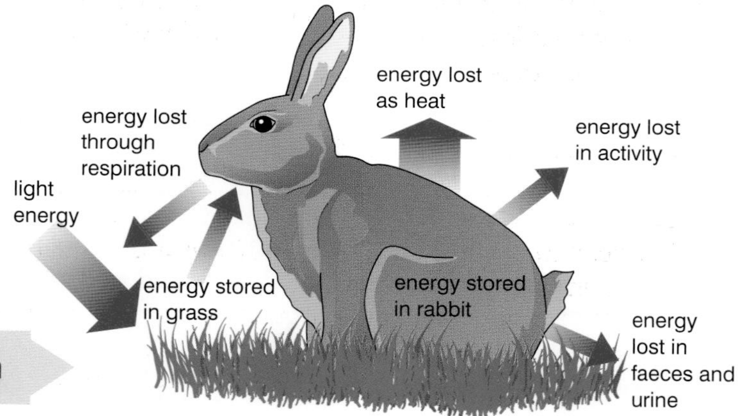

How energy is lost in a food chain

a Give three ways that energy is lost in food chains.

b Why do food chains usually contain a maximum of five different organisms?

Some pyramids of numbers are a strange shape like this one based on a tree:

top carnivores
intermediate carnivores
herbivores
producers

A pyramid starting with a tree

One tree supports many herbivores so the 'pyramid' is not pyramid shaped!

A pyramid ending with a parasite looks like this:

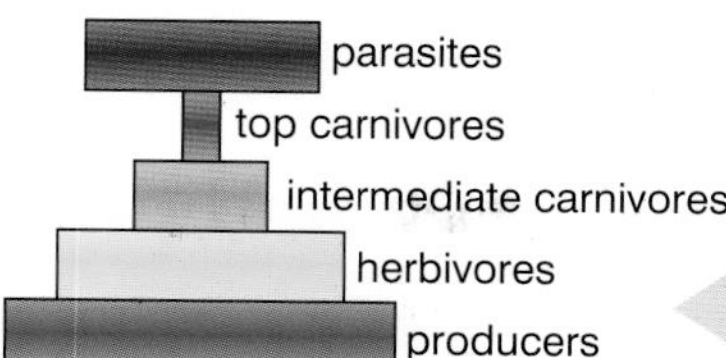

A pyramid ending with a parasite

Lots of small parasites live on one **host** animal.

Bioaccumulation

DDT is a **pesticide**, a chemical used to kill pests. In the 1940s DDT was sprayed on a lake in California to kill mosquitoes. Some of the DDT got into tiny plants:

tiny plant ⟶ small animals ⟶ fish ⟶ grebe (water bird)

Each tiny plant only has a small amount of DDT (1 ppm means 1 part per million – for every gram of tiny plant there is 1 millionth of a gram of DDT). However each small animal eats many plankton. DDT does not break down easily and living things cannot get rid of it easily. This is described as persistent.

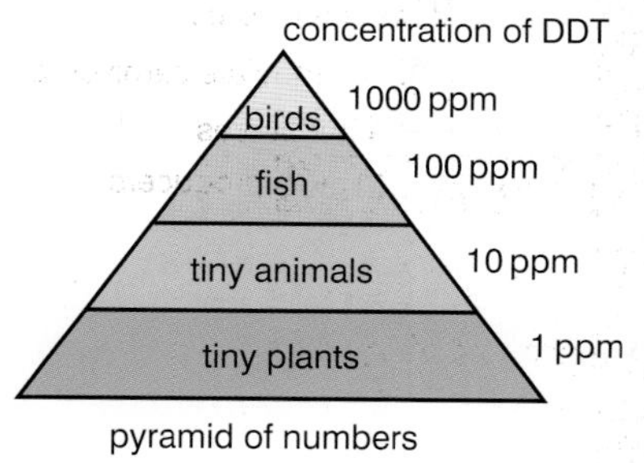

The amount of DDT becomes more concentrated in the small animals. This is called **bioaccumulation**. Fish eat lots of small animals and grebes eat lots of fish. The DDT concentration in grebes is so high that they get poisoned.

c How did DDT get into this food chain?

d What happens to the concentration of DDT as it passes up the food chain?

Summary Questions

1. Match the food chains to the correct pyramid of numbers:
   a) grass ⟶ crane fly larva ⟶ swallow
   b) rose bush ⟶ greenfly ⟶ blue tits

   i) ii) iii)

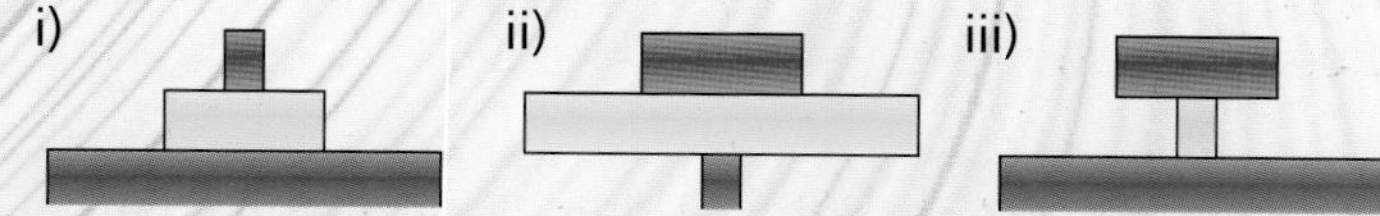

2. Draw a flow chart or diagram to explain how grebes were poisoned by DDT.
3. Look at these two food chains:
   corn ⟶ human
   corn ⟶ pig ⟶ human
   Which one provides the most energy for humans? Explain your answer.

KEY WORDS
pyramids of numbers
pesticide
bioaccumulation

Predators and Prey

B2.8

- How are predators adapted to catch prey?
- How are plants and prey adapted to avoid being eaten?
- How do populations of predators and prey affect each other?

What are predators and prey?

Predators eat other animals. Animals that get hunted are called **prey**.

Predators are adapted to find, catch and eat other animals.

A Bengal tiger has:

- striped fur to camouflage it
- powerful legs to run fast
- forward-facing eyes to see prey
- sharp claws to grab and hold prey
- sharp teeth to grip prey and tear flesh.

A Bengal tiger, a predator

A giraffe, feeding on acacia leaves, is not a predator

Prey animals are adapted to avoid being eaten.

A rabbit has:

- large ears to hear danger
- powerful legs to run away from predators
- eyes on the side of the head to give 360° vision.

A rabbit, prey

Many plants have adaptations that help them to avoid being eaten.

Acacia trees start to produce a bitter chemical when giraffes eat their leaves. Giraffes move to another tree to find one that has not started to produce the chemical!

Help Yourself

Remember that animals that eat plants are **not** *predators and plants are* **not** *their prey.*

a Describe adaptations of the golden eagle that make it an effective predator.

Golden eagle

b Describe how the wildebeest is adapted to avoid being eaten by predators such as lions and hyenas.

Wildebeest

Did You Know?

Acacias make ethene gas when animals eat their leaves. This is carried on the wind to warn other acacias to start making the bitter substance! Giraffes learn always to move to trees upwind of the one they are eating!

Canadian lynx and snowshoe hare

activity

Predators and prey

Find photographs or make drawings of a predator and its prey. Do not choose animals shown on this page.

Use them to make a poster. Show how the predator is adapted to find, catch and kill its prey. Show how the prey is adapted to avoid being eaten.

Predator–prey relationships

The size of the populations of predators and prey are closely linked. If there are lots of prey, the predator population will increase. If there are lots of predators, the prey population will fall.

The Hudson's Bay Company of Canada has records of the numbers of animal skins bought over a period of about 100 years. We can use these records to examine the relationship between the number of snowshoe hares and their main predator, the Canadian lynx.

The graph shows that when there are lots of hares, the lynx have plenty to eat and so breed more. When there are more lynx the hare population goes down. This means there is less food for the lynx so some will starve and the population goes down. A small lynx population means there will be more hares … and so it goes on!

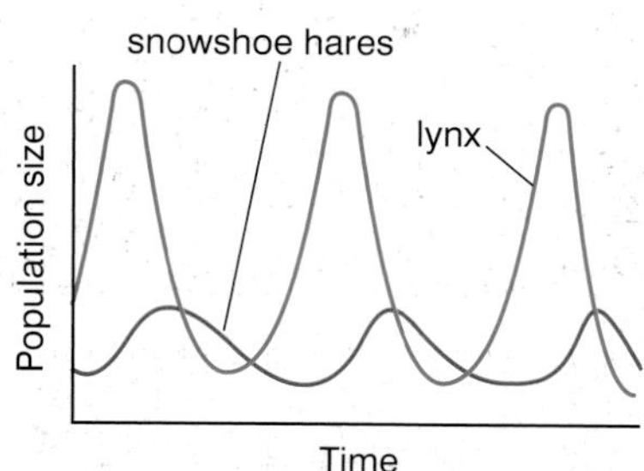

Link up to…

MATHEMATICS

Biologists use mathematical and computer skills to make 'models' of predator–prey relationships.

Summary Questions

1. Draw a picture of a 'super-predator'. Label your drawing.
2. Draw a super-prey that cannot be caught by a predator.
3. Draw and label a graph to explain the relationship between the numbers of snowshoe hares and lynx.

KEY WORDS

predator
prey

Habitats and Adaptation

- What is a habitat?
- How are animals and plants adapted to live in a particular habitat?

Where do they live?

The place where organisms live is called a **habitat**. A habitat must provide the food, water, shelter and oxygen that animals need to survive. It must also provide light, carbon dioxide and minerals for plants.

There are lots of different types of habitat. Different habitats have different environmental conditions.

Desert

Arctic

Rainforest

Sea shore

(a) Look at the habitats in the photographs. For each one, describe the environmental conditions.

(b) What factors could make it difficult for animals and plants to survive in each habitat?

(c) Name some animals and plants that might live in each habitat.

Animals and plants are adapted to live in a particular habitat.

The arctic fox has:

- small ears to reduce heat loss
- white fur to provide camouflage against snow
- thick fur to keep warm.

The desert fox has:

- large ears to lose heat
- sandy coloured fur to provide camouflage against the sand
- thin fur to keep cool.

Desert fox

Arctic fox

Both of these species of fox eat similar food. They have similar adaptations to help them to catch food. They live in very different habitats and have special adaptations that help them to survive.

The cactus has:

- spines so that animals will not try to eat the plant to get water
- no leaves so that water is not lost through them
- thick stems that store water
- roots that go deep into the ground to reach water.

Cactus plants

activity

Adaptations for a habitat

Choose three habitats.

For each habitat find a photograph or draw a picture of an animal living there.

Label the picture to identify the adaptations that help the animal to survive in the habitat. Do not choose animals shown on this page.

Summary Questions

1. For each of the photographs below, describe:
   a) Where the organism lives and what it eats.
   b) How it is adapted to survive in its habitat.

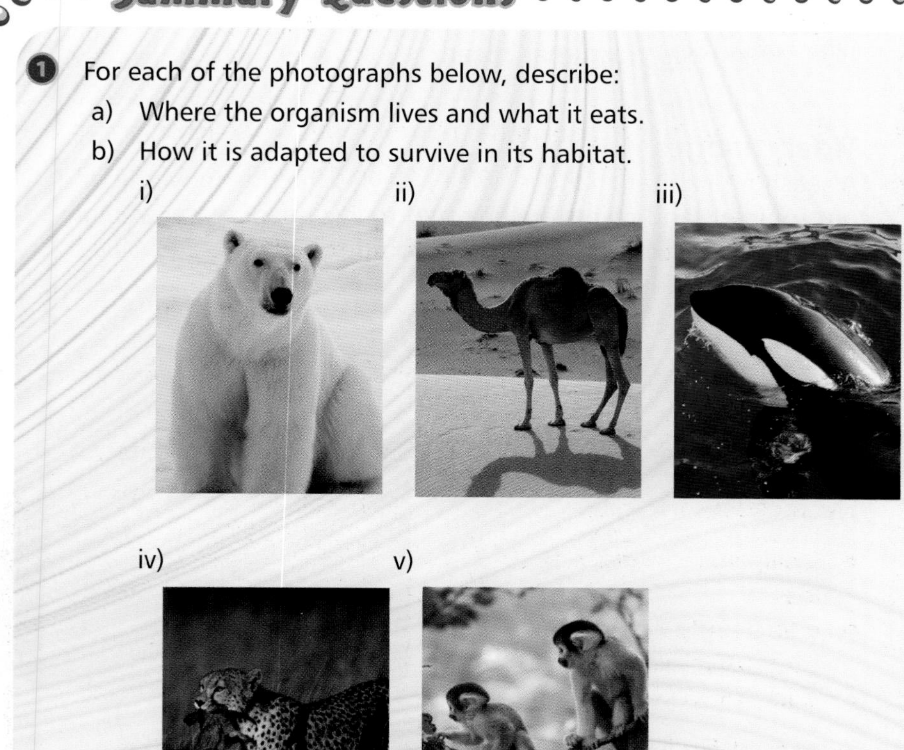

KEY WORDS

habitat

Habitats Changing

B2.10

- How do habitats change daily?
- How do habitats change with the seasons?
- How do animals and plants cope with these changes?

Night and day

Habitats change through every 24 hours. During the day it is light and warmer than at night. At night it is dark and colder. Some animals are active during the day, using their eyesight to find food. Amphibians and reptiles are cold blooded and are active during the day when it is warmer.

Other animals are **nocturnal**, which means they are active at night. Some animals, like dormice, are nocturnal so that they are less likely to be eaten. However, there are also nocturnal predators, such as owls, which prey on them.

Owls' large eyes help them to see their prey at night. Their flat faces help to funnel the sound of prey to their ears.

Some plants have flowers that open at night. They are pollinated by moths and other nocturnal creatures. These flowers produce strong scents so that moths can find them.

Moths' large antennae detect smells. They use them to find flowers. Males can also smell females over long distances!

Barn owl

Moth antennae

a What does 'nocturnal' mean?

Tidal changes

A seashore habitat changes twice a day as the tide moves in and out. Animals and plants are adapted to live under water at high tide and to prevent drying out at low tide.

Sea anemone – tide in

Sea anemone – tide out

b Sea anemones live in rock pools. How are they adapted to survive tidal changes?

Seasonal changes

Environmental conditions change throughout the year. These are called seasonal changes. Some trees, called **deciduous** trees, lose their leaves in the winter because there is not much light and it is too cold for photosynthesis. If they kept their leaves, the weight of snow could damage the trees.

Evergreen trees do not lose their leaves. Snow just slides off their needle-like leaves.

Some animals have special seasonal adaptations. Stoats in cold places grow white fur in the winter.

Seasonal changes in a tree

Stoat in summer

Stoat in winter

In its winter coat the stoat is known as the 'ermine'.

Did You Know?

The fur of the ermine was used to trim the robes of members of the House of Lords. Fake fur is used now.

Migration

Many species of birds **migrate**. Some, such as swallows, spend the summer in Europe and migrate to Africa in winter. Others, such as species of ducks and geese, spend the summer in the Arctic and migrate to Europe in the winter. Arctic terns fly about 40 000 km each year, flying from the Arctic to the Antarctic and back!

Huge herds of wildebeest migrate across the plains of Africa, following the rain.

Migrating wildebeest wait to cross a river

Hibernation

Some animals eat lots of food in the summer to build up fat reserves. They then slow down their body activity and **hibernate** through the winter. Hedgehogs, dormice, ladybirds and frogs all hibernate.

Summary Questions

1. Make a poster to explain how animals are adapted for either a) daily changes or b) seasonal changes in environmental conditions.

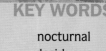

KEY WORDS

nocturnal
deciduous
migrate
hibernate

Investigating a Habitat (1)

- How can we use keys to help us to identify living things?
- How can we collect animals and plants in a habitat?

A pooter

A pitfall trap

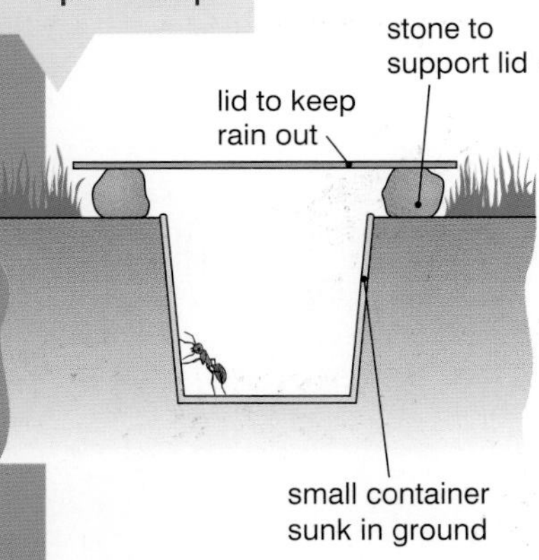

Tree beating

What lives in a habitat?

You can observe large mammals and birds by watching quietly, especially if you cannot easily be seen. You could set up a video camera to record when you are not there. Play it back at high speed and watch out for animals!

To find smaller animals there are several different methods. Whichever methods you use, be careful not to damage the habitat or harm any animals.

- A **sweep net** is a large net that you sweep through long grass. You can then examine what you find.
- A **pooter** is a transparent container with two tubes. Put one tube near a creature and suck. You will suck small animals into the container. A filter stops you from swallowing any animals but make sure you suck on the right tube!
- A **pitfall trap** is a container buried in the ground and left there. Small animals fall into the container. Make sure you cover it with a stone or loose lid to keep rain out.
- **Tree beating** involves holding a white sheet or a large sheet of card under a tree or bush. Gently hit or shake the bush and collect the animals that fall onto the sheet or card.

activity

Studying a habitat

- Choose a habitat to study.
- Use the most appropriate methods of collecting animals that live there.
- Try to identify the organisms you find.
- Count the numbers of each species.
- Use graphs, tables, pie charts, etc. to present your findings.
- Try to explain how each organism is adapted to the place you found it, for example:
  - Does it have wings? What shape is it? What colour is it?
- Research the species you have identified. What does it eat? What eats it? Use your findings to make a food web for your habitat.

 **Safety:** Take care when working outside.

Identification keys

You may not recognise the animals and plants you find. You could spend hours looking in books or on the Internet to find a particular organism or you could use an **identification key**.

A key has a number of questions or statements. Start at the beginning and the key will lead you to the animal or plant you want to identify. There are two types of key:

- a branching key
- a dichotomous key

A branching key

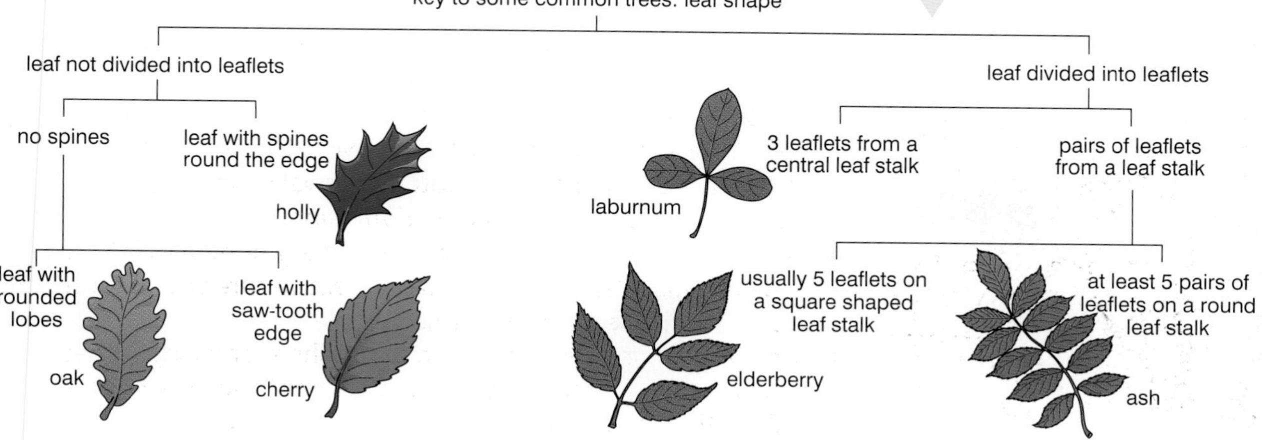

A dichotomous key

a Use this key to identify the grassland invertebrates shown to the right:

1 Does it have legs?
Yes – go to 2 No – snail

2 Does it have 3 or 4 pairs of legs?
Yes – go to 3 No – centipede

3 Does it have 3 pairs of legs?
Yes – go to 4 No – spider

4 Does it have spots on its body?
Yes – ladybird No – ground beetle

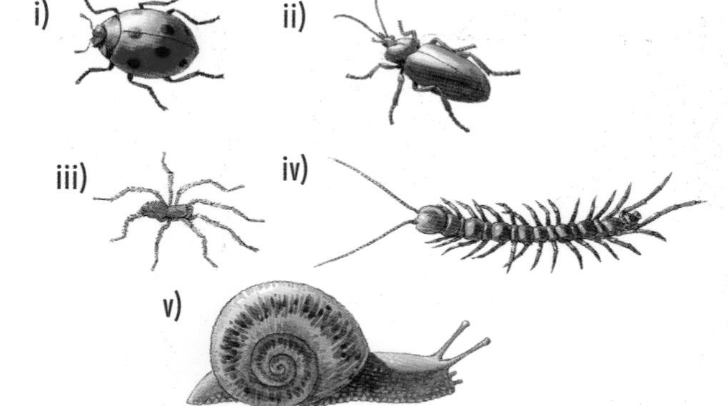

Summary Questions

1. Make a key so that an alien could identify the following: knife, fork, spoon, cup, plate, bowl.
2. Make a poster to display what you found out about the habitat you studied.

KEY WORDS

sweep net
pooter
pitfall trap
tree beating
identification key

Investigating a Habitat (2)

B2.12

- How can we measure environmental factors in a habitat?
- How can we estimate the number of plants in a habitat?

Sampling

It would take a long time to count every single plant in a habitat. We can estimate the number of plants by **sampling**. We use a **quadrat** to make sure that our samples are the same size.

A quadrat is a square, either 1 m² or 0.25 m². We take random samples from different parts of the area we are investigating.

Some pupils are discussing how to estimate the number of daisies on Kiran's lawn

Don't put the quadrat there – There are no daisies there.
I'm not doing it over there either – it's too muddy!
That looks like a good spot – There are loads of daisies there.

ⓐ What is meant by a 'random' sample?

ⓑ Why is it important that our samples are random?

ⓒ Why wouldn't it be useful to use the first four digits of a phone number?

Selecting a random sample: Kiran's phone number ends in 5827

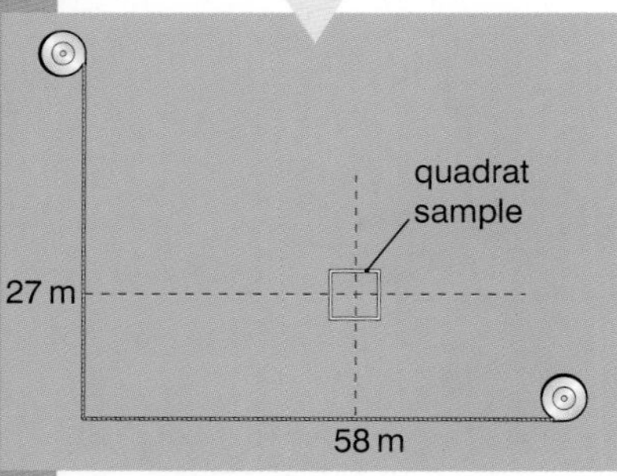
quadrat sample
27 m
58 m

activity

Estimating by sampling

We can use a quadrat to estimate the number of dandelions on a football pitch or similar area.

- Take ten random samples and count the number of dandelions in each.
- Work out the average per m².
- Work out the area of the football pitch (width × length).
- Work out the total number of dandelions by (area × average number per m²).

Safety: Take care when working outside.

There are daisies in 5 squares so the percentage cover is 20%

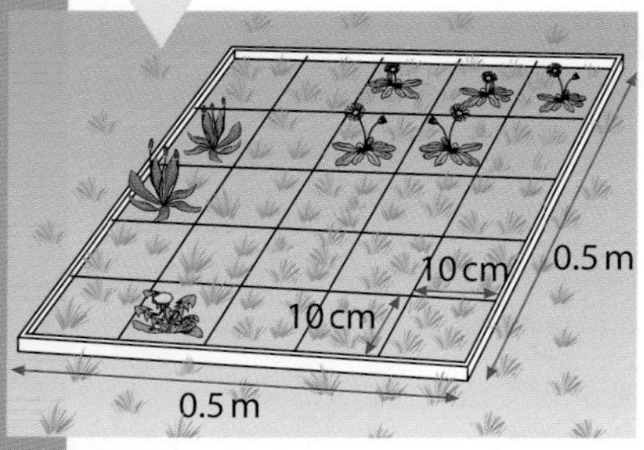
10 cm
10 cm
0.5 m
0.5 m

Another way of sampling is to estimate the percentage cover in each quadrat. Some quadrats have wire dividing them into 25 smaller squares. If yours doesn't you can estimate percentage cover by eye.

Transects

A transect is a useful way of finding out how a plant population changes with different conditions.

activity

Taking a transect

- Mark out a transect with a long piece of string.
- Use your quadrat to estimate the percentage cover of plants at intervals.
- At the same intervals, measure **physical factors** such as light intensity, soil temperature and air temperature.
- You could also measure soil pH and soil moisture. Your teacher will show you how.
- Record your results in a table like this:

A transect

Distance (m)	Percentage cover			
	Grass	Daisy	Dandelion	Clover
0	80	10	10	0
2.5	70	10	10	10
5	70	10	20	0
7.5	60	10	20	10
10	50	20	20	10

- Use your results to draw a **kite diagram** like the one right. The diagram shows the greater the percentage cover, the wider the bars in the kite diagram.
- Add information about physical factors to your kite diagram.
- How do the physical factors you measured affect the percentage cover of different plants?

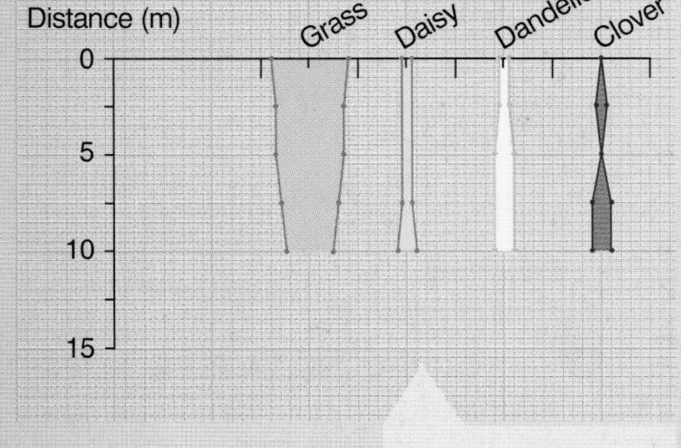

A kite diagram

Safety: Take care when working outside.

Summary Questions

1. Carlo and Joe used a 0.5 × 0.5 m quadrat to estimate the number of daisies on Carlo's croquet lawn measuring 20 m × 10 m. The ground was muddy so they did not walk too far onto the lawn. The diagram shows where they put their quadrats.

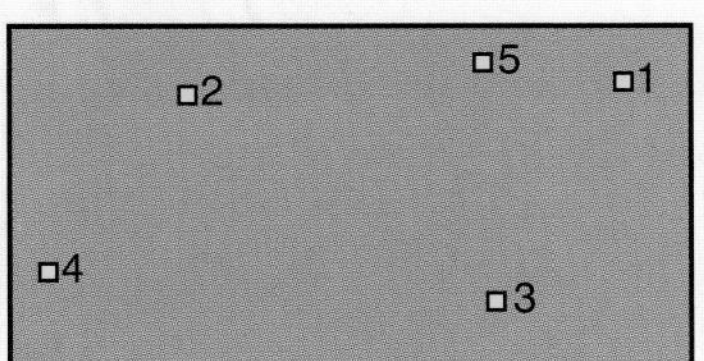

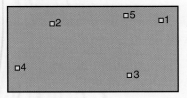

Their results are shown below:

Quadrat	1	2	3	4	5
Number of daisies	3	8	6	2	5

Estimate the total number of daisies on the lawn.
How could they make their estimate more reliable?

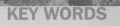

KEY WORDS

sampling
quadrat
physical factor
kite diagram

know your stuff

▼ Question 1 (level 4)

The diagram below shows part of a farmland food web.

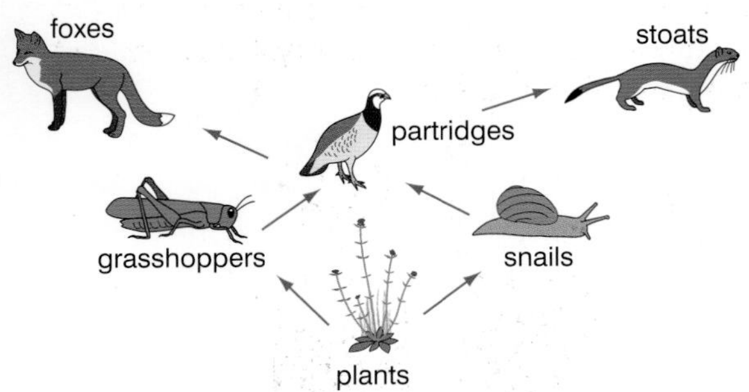

a Name *one* predator and its prey in the food web. [1]

b Why are the plants in the food web called 'producers'?
Choose the correct answer from the list below:

They lose their leaves in the autumn.

They make food by photosynthesis.

They have long roots.

They have small flowers. [1]

c Partridges lay their eggs in nests on the ground.

(i) The eggs are the same colour as the ground.
How does this help partridges to survive? [1]

(ii) Why could laying eggs on the ground result in fewer partridge chicks? [1]

▼ Question 2 (level 4)

The animals shown below live in different parts of a river.

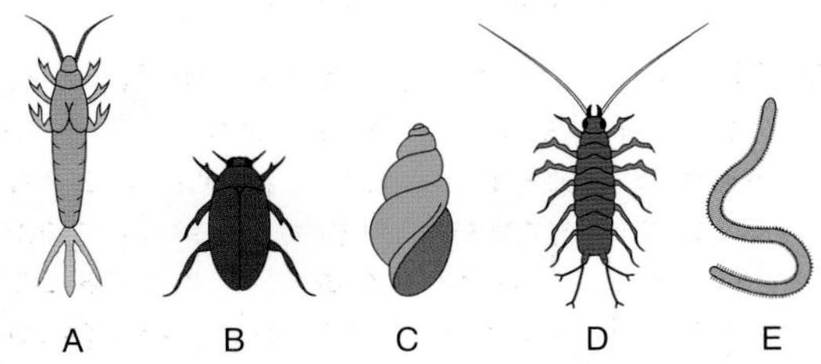

Use this key to identify animals A, B and C. [3]

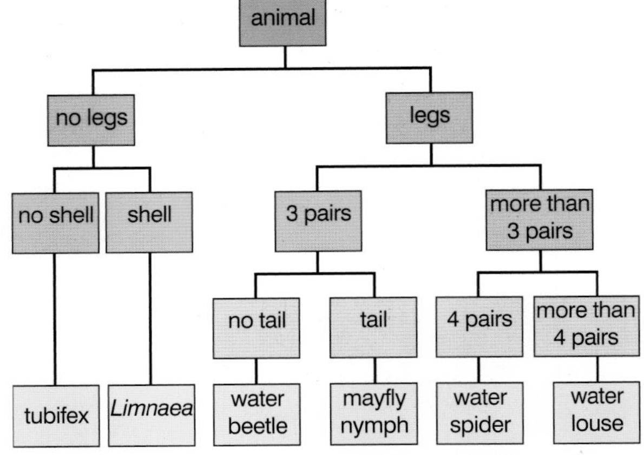

▼ Question 3 (level 6)

The diagram shows four living things.

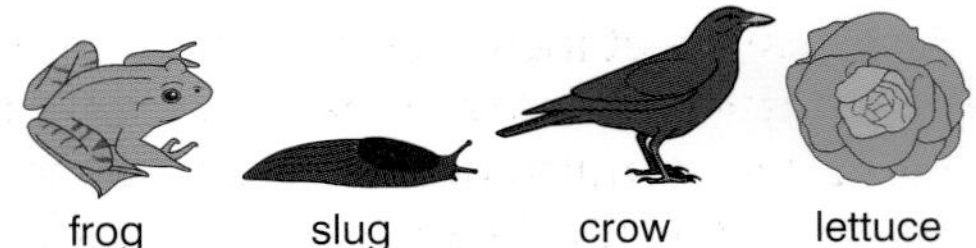

Frogs eat slugs. Slugs eat lettuce leaves. Crows eat frogs.

a Write down the food chain which has crows, frogs, lettuce and slugs in it. [1]

b Name the producer in the food chain. [1]

c From the living things shown above, give the name of a predator and its prey. [2]

d Lynx are wild cats that feed on snowshoe hares. The graph shows how the population of both animals varies over time:

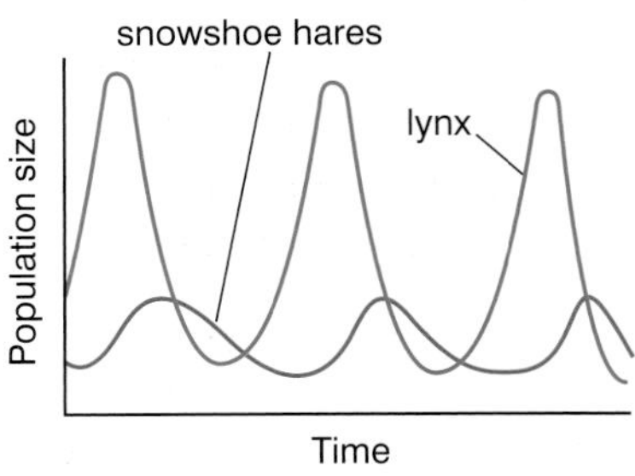

Explain the graph. [4]

How Science Works

▼ Question 1 (level 6)

Manpreet did an investigation to find out how temperature affects the number of bubbles produced by waterweed in one minute. Her apparatus is shown below.

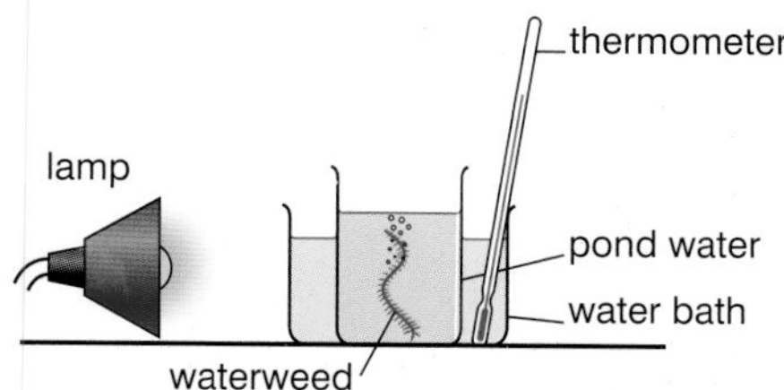

When the temperature of the water was 10°C the waterweed did *not* produce bubbles.

a The waterweed started to produce bubbles when Manpreet increased the temperature of the water in the water bath to 20°C. She waited two minutes before counting the bubbles. Why did she wait two minutes before she started to count the bubbles? [1]

b Manpreet counted the number of bubbles made at six different temperatures. Here is a table of her results:

Temperature of water bath (°C)	10	20	30	40	50	60
Number of bubbles produced per minute	0	8	12	12	8	2

(i) Draw a graph of Manpreet's results. [4]

(ii) Draw a smooth curve on the graph. [1]

(iii) At what temperature were the most bubbles produced per minute? [1]

c Manpreet made a prediction that the higher the temperature the more bubbles would be produced. Which parts of the graph support Manpreet's prediction? [1]

d Manpreet's results do *not* show the exact temperature at which most bubbles were produced. How could she improve the results she collects so she can find this temperature? [1]

▼ Question 2 (level 7)

In the 17th century a Belgian scientist, Van Helmont, planted a small tree in a tub of dry soil. For the next five years he watered the plant with rain water when it got dry. He did not add anything else to the tub.

After five years he took the tree from the tub and weighed it. He also dried and weighed the soil. Here is a table of his results:

	Mass of willow tree (kg)	**Mass of dried soil (kg)**
At the start	2.3	90.6
Five years later	76.7	90.5

a Van Helmont reached the conclusion that the increase in mass of the willow tree was due to a gain in water.

(i) What *two* pieces of evidence led Van Helmont to this conclusion? [2]

(ii) We now know that Van Helmont's conclusion is *not* correct. Explain why the mass of the willow tree increased. [2]

b Van Helmont thought that a plant would always grow faster if it was given more water. We now know that this is *not* true. Give *two* environmental conditions which can slow down the growth of a plant, even when it has plenty of water. [2]

c The **fresh mass** of a plant includes water. To measure plant growth accurately, scientists work out the increase in **dry mass** rather than the increase in fresh mass of a plant. Why is finding the increase in dry mass a more reliable way to measure plant growth? [1]

Elements and Compounds

Different materials

Look around your classroom; how many different materials can you see? A 'material' is the word that scientists use to describe what objects are made from. Cotton is a material and so is air.

We use the word 'material' or 'substance' for anything made from atoms or molecules

It is important to know the properties of a material, so that you can choose a job for it. Copper is a good electrical conductor (lets electricity flow easily through it). This property makes it useful for electrical wires.

Did You Know?

Sometimes designer materials are made before a use can be found. Dr Silver at 3M made a new sticky adhesive in 1968, but it wasn't until many years later that a use was found for it – sticky notes!

a) What property makes steel useful for construction?

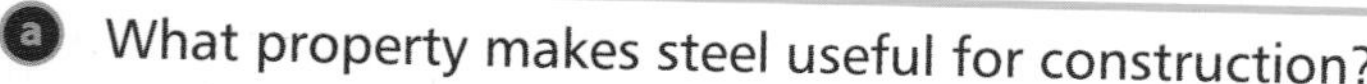

b) What material would you use to make a magnet and why?

Scientists have researched into materials and they have found out how to make them. Today, materials can be designed for a specific job, rather than choosing from a list of materials and 'making do'.

Some materials are natural, for example, clay. Natural materials may be non-renewable (will run out) and dug out of the ground from quarries or mines. Other natural materials are renewable (will not run out) and are made from animals or plants.

The opposite of a natural material is a man-made, or synthetic, material. These are materials that scientists have made in the lab. Plastics are an example of a synthetic material. Crude oil is a natural material that is changed by chemical reactions to make plastics.

Why might a hiker choose a synthetic jacket made from Thinsulate rather than a natural woollen jacket?

Some materials are pure and may be the elements listed in the Periodic Table. However most materials are mixtures of more than one substance, not chemically joined together.

activity

Classifying materials

Designers need to know the properties of different materials so that they can choose the best one for a job. If there is not a material suitable for the job, they could write a 'wish-list' of properties. Then a team of scientists can develop materials just for that specific job. There are data bases of materials that list their properties, so designers can quickly search to find a short list of materials that may be suitable for the job.

You will be given a selection of different materials. You should classify them by their state of matter and whether they are natural or synthetic.

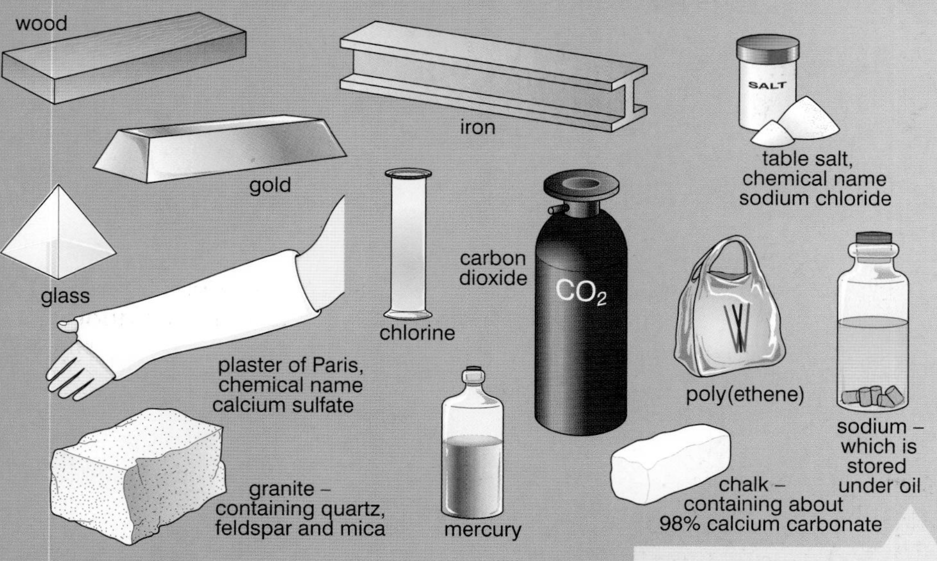

Classify the different materials

Copy and complete the table below:

Material	State	Type

Using the Internet, find out the main properties for each material that you have studied and one of its uses. Can you say which property makes the material useful for that job?

Building Blocks

- What is matter made from?
- What is an element?

Did You Know?

Over 5000 new materials are registered by scientists every day!
I think you could register that as a new substance.

What's the matter?

There are billions of different **materials** in the Universe and they are all made up of matter. **Atoms** are the particles that make up all matter.

There are only about 100 types of atom. The materials that are made up of only one type of atom are called **elements**. All of the elements are listed in the **Periodic Table** (see below). When these atoms join or mix in different ways we get all the different materials that we know and maybe some we have yet to discover!

One model for thinking about the elements is to imagine that each type of atom is like a letter in the alphabet. Let's imagine that we have the letters H, O and P. You can arrange them in pairs in six different ways. This means that you would have six different materials. Each material would have different properties and so different uses.

a Write the six different ways to arrange the letters H, O and P in pairs.

b One of the drawbacks of this model is that there are over 100 elements, not just three. What other drawbacks can you think of?

H Hydrogen																	He Helium
Li Lithium	Be Beryllium											B Boron	C Carbon	N Nitrogen	O Oxygen	F Fluorine	Ne Neon
Na Sodium	Mg Magnesium											Al Aluminium	Si Silicon	P Phosphorus	S Sulfur	Cl Chlorine	Ar Argon
K Potassium	Ca Calcium	Sc Scandium	Ti Titanium	V Vanadium	Cr Chromium	Mn Manganese	Fe Iron	Co Cobalt	Ni Nickel	Cu Copper	Zn Zinc	Ga Gallium	Ge Germanium	As Arsenic	Se Selenium	Br Bromine	Kr Krypton
Rb Rubidium	Sr Strontium	Y Yttrium	Zr Zirconium	Nb Niobium	Mo Molybdenum	Tc Technetium	Ru Ruthenium	Rh Rhodium	Pd Palladium	Ag Silver	Cd Cadmium	In Indium	Sn Tin	Sb Antimony	Te Tellurium	I Iodine	Xe Xenon
Cs Caesium	Ba Barium	La ▸ Lanthanum	Hf Hafnium	Ta Tantalum	W Tungsten	Re Rhenium	Os Osmium	Ir Iridium	Pt Platinum	Au Gold	Hg Mercury	Tl Thallium	Pb Lead	Bi Bismuth	Po Polonium	At Astatine	Rn Radon
Fr Francium	Ra Radium	Ac ▸▸ Actinium															

▸	Ce Cerium	Pr Praseodymium	Nd Neodymium	Pm Promethium	Sm Samarium	Eu Europium	Gd Gadolinium	Tb Terbium	Dy Dysprosium	Ho Holmium	Er Erbium	Tm Thulium	Yb Ytterbium	Lu Lutetium
▸▸	Th Thorium	Pa Protactinium	U Uranium	Np Neptunium	Pu Plutonium	Am Americium	Cm Curium	Bk Berkelium	Cf Californium	Es Einsteinium	Fm Fermium	Md Mendelevium	No Nobelium	Lr Lawrencium

The Periodic Table

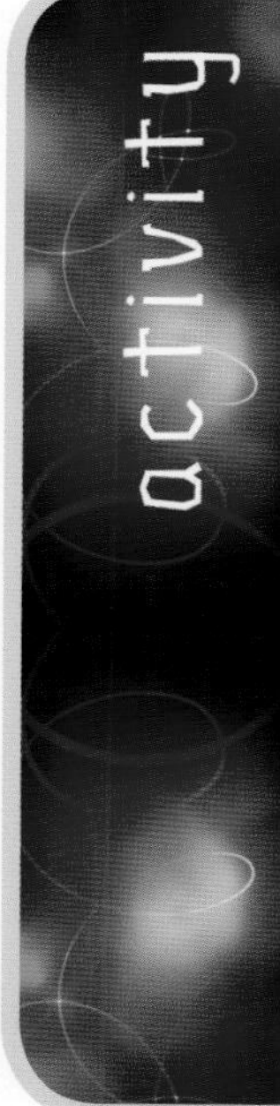

Making a model of elements

You are going to make different models for elements. You will list the good points of each model and the drawbacks. Use each modelling substance to make a model of the particles in an element and a mixture of two elements. How might you display your ideas?

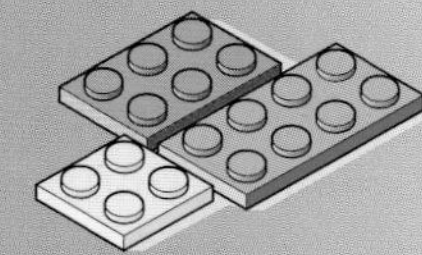

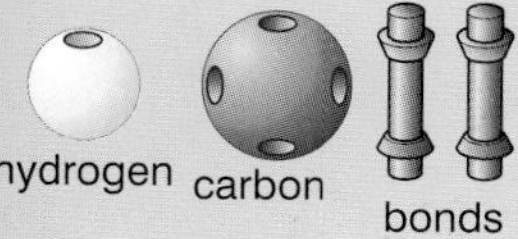

- Use Plasticine to make a model.
- Use Lego bricks to make a model.
- Use molecular model kits to make a model.

Which model do you think is best and why? Can you think of a different model?

Models are an important way that scientists can explain observations and make predictions. All models simplify what is really happening. This means that they have drawbacks. Weather forecasters use models of how the atmosphere works to make predictions about the weather – but they can still get it very wrong sometimes.

Did You Know?

Steel is a construction material made from a mixture of mainly iron and other elements. If you change the percentage of carbon in the steel, it changes from being malleable (bends easily) to brittle (breaks easily). This gives the steels different uses. Different amounts of the same elements in a mixture can change a material's properties.

Summary Questions

1. Match up the key words with their definitions:

a) material	contains only one type of atom
b) atom	a list of all the elements
c) element	something that is made up of the same or different atoms
d) Periodic Table	the smallest building block of matter

2. Use the Periodic Table to help you answer the following questions:
   a) Which element do we breathe from the air to keep us alive?
   b) Which elements do we use to make jewellery?
   c) Which element can we test with a lighted splint and hear a pop?

3. The ancient Greeks were the first to come up with the idea that everything was made up of small particles, which could not be split. They called these particles 'atoms'. However, the atom contains small particles that cannot exist on their own. Do some research to find out what an atom is made from and who discovered each particle. Key names to use in an Internet search would be: J.J. Thomson, Ernest Rutherford, Niels Bohr and James Chadwick.

KEY WORDS

material
atom
element
Periodic Table

Elements and their Properties

- What are the properties of some elements?
- How can I classify elements?

It's elementary!

Elements are made up of only one type of atom. There are only about 100 elements that are naturally found on Earth. When nuclear reactors and weapons were invented, other heavier elements were found. All of the elements, whether they are natural or man-made, are listed in the Periodic Table (see page 56).

a How many naturally occurring elements are there?

Stretch Yourself

The newly discovered heavy elements, like ununtrium (element 113) are very unstable. Their atoms are so big and heavy that they survive for only milliseconds before breaking into smaller atoms. Not much is known about the properties of these elements. Why do you think this is?

Each element has different properties

We can classify elements in many different ways. The main way of classifying elements is as a **metal** or a **non-metal**.

b How else could elements be classified?

science@work

Medical professionals sometimes use radioactive elements to get clear images of patients' organs. Radioactive iodine can be injected into patients with thyroid problems. The iodine makes its way to the thyroid gland in the patient's neck. Iodine shows clearly on a scan, so doctors can get a clearer picture of the patient's thyroid gland.

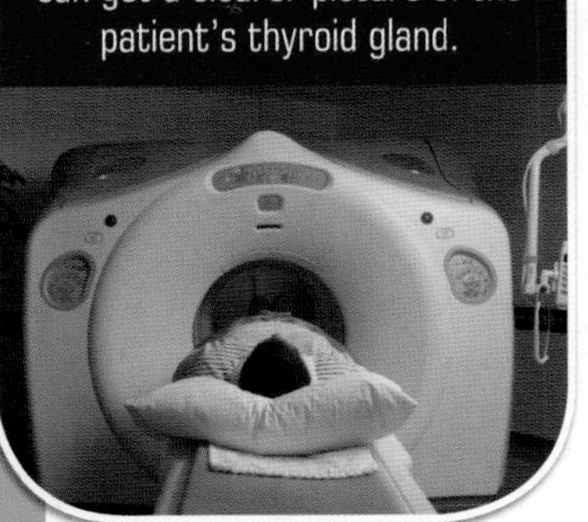

Studying elements

Very unreactive elements, such as gold, have been known to humans for thousands of years. More reactive elements like sodium have only been discovered since electricity could be used. Scientists are very interested in elements and research is still happening today to find out if there are any more that we do not know about.

You will be given a selection of elements.

You should describe what they look like and classify them as metal or non-metal. You could even make a prediction about their **properties** based on their classification.

c What are the general properties of a metal and a non-metal?

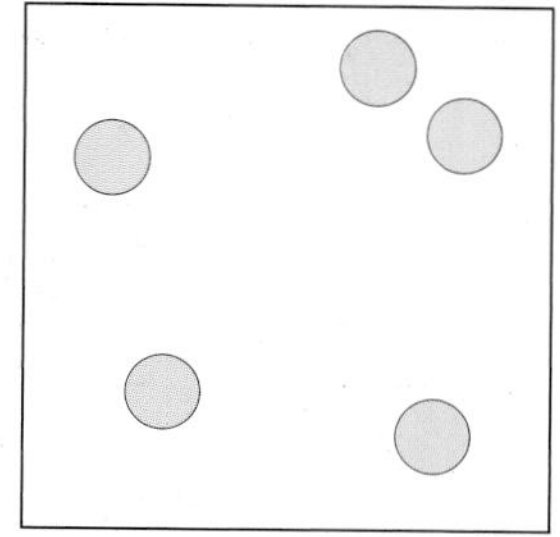

A particle diagram of helium gas

We can draw particle diagrams to show an element. All of the particles should be the same – same colour and same size. This is because all of the atoms in an element are the same.

Atoms are very tiny particles. The average atom is about 1/5 000 000 mm across! That means about 15 million atoms on top of each other would be as high as a pound coin!

You can't see atoms using just your eyes, or even with a school microscope. A special machine called an electron microscope can be used to look at the shapes of atoms.

No, we can't see atoms through our microscope ...atoms are much too small for that! Even a tiny plant cell contains millions of atoms.

Great Debates

Scientists get money from lots of different places, including governments and companies. Is it important that scientists are allowed to try to find new elements, when we know very little about some elements like francium (Fr)? Do you think that taxpayers' money should be used to help find new elements?

Summary Questions

1. Copy and complete the passage using the words below:

   elements metals non-metals bromine atom mercury

   Elements are made up of only one type of .... At room temperature most elements are solid .... The only liquid metal at room temperature is .... All of the elements that are gases at room temperature are also .... The only liquid non-metal at room temperature is .... There are about three times more metal elements than non-metal ....

2. Draw a particle diagram to show a solid element.

3. Choose an element and find out about it! Who discovered it, and when? What are its properties and its uses? Make a poster to show your research.

KEY WORDS

element
metal
non-metal
properties

Symbols

- Why do elements have symbols?
- How do I find out the symbols of elements?

In your element

In about 400 BC, the Greeks suggested that everything was made out of small particles that could not be broken up. Democritus was a student of Leucippus and it is difficult to know which Greek suggested the idea of atoms. However, Democritus is often credited with calling the particles atoma – which means indivisible units. This the root of the English word 'atom'.

John Dalton, an English scientist, was the first to make a list of all the materials which he believed were elements. He based his theories on experiments, but did include some substances which we now know are not elements. He listed his elements in order of atomic weight and gave each one a **symbol**.

ELEMENTS

Element	W[t]	Element	W[t]
Hydrogen	1	Strontian	46
Azote	5	Barytes	68
Carbon	5.4	Iron	50
Oxygen	7	Zinc	56
Phosphorus	9	Copper	56
Sulphur	13	Lead	90
Magnesia	20	Silver	190
Lime	24	Gold	190
Soda	28	Platina	190
Potash	42	Mercury	167

Dalton's table of elements in 1805

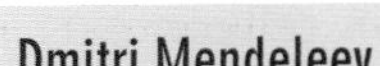

Dmitri Mendeleev

Did You Know?

John Dalton was born in 1766 and died in 1844. He spent most of his life teaching in Manchester. Dalton suggested his atomic theory in 1805 and based his thoughts on experiment. His ideas were similar to those of Democritus, who suggested in 400 BC that things were made of particles.

John Dalton

a Look carefully at Dalton's list of elements. Name one chemical on the list that is actually not an element.

Dmitri Mendeleev, a Russian scientist, created the first **Periodic Table** in the 19th century. In 1869, when he published his table, many more elements were known. He listed them mainly in order of their atomic weights. He made sure that elements with similar chemical properties lined up. He also left gaps for elements that he predicted were yet to be discovered.

b Look carefully at the Periodic Table. What is the name and symbol of the element named after Mendeleev?

Stretch Yourself

Dmitri Mendeleev was from a poor Russian family. His mother worked very hard to get her son into a local university. When Mendeleev put forward his Periodic Table, many scientists refused to accept his ideas. Why do you think this was?

H Hydrogen																	He Helium
Li Lithium	Be Beryllium											B Boron	C Carbon	N Nitrogen	O Oxygen	F Fluorine	Ne Neon
Na Sodium	Mg Magnesium											Al Aluminium	Si Silicon	P Phosphorus	S Sulfur	Cl Chlorine	Ar Argon
K Potassium	Ca Calcium	Sc Scandium	Ti Titanium	V Vanadium	Cr Chromium	Mn Manganese	Fe Iron	Co Cobalt	Ni Nickel	Cu Copper	Zn Zinc	Ga Gallium	Ge Germanium	As Arsenic	Se Selenium	Br Bromine	Kr Krypton
Rb Rubidium	Sr Strontium	Y Yttrium	Zr Zirconium	Nb Niobium	Mo Molybdenum	Tc Technetium	Ru Ruthenium	Rh Rhodium	Pd Palladium	Ag Silver	Cd Cadmium	In Indium	Sn Tin	Sb Antimony	Te Tellurium	I Iodine	Xe Xenon
Cs Caesium	Ba Barium	La ▸ Lanthanum	Hf Hafnium	Ta Tantalum	W Tungsten	Re Rhenium	Os Osmium	Ir Iridium	Pt Platinum	Au Gold	Hg Mercury	Tl Thallium	Pb Lead	Bi Bismuth	Po Polonium	At Astatine	Rn Radon
Fr Francium	Ra Radium	Ac ▸▸ Actinium															

▸	Ce Cerium	Pr Praseodymium	Nd Neodymium	Pm Promethium	Sm Samarium	Eu Europium	Gd Gadolinium	Tb Terbium	Dy Dysprosium	Ho Holmium	Er Erbium	Tm Thulium	Yb Ytterbium	Lu Lutetium
▸▸	Th Thorium	Pa Protactinium	U Uranium	Np Neptunium	Pu Plutonium	Am Americium	Cm Curium	Bk Berkelium	Cf Californium	Es Einsteinium	Fm Fermium	Md Mendelevium	No Nobelium	Lr Lawrencium

The Periodic Table

In the Periodic Table, each element has its own box with information about its atoms. Mendeleev also gave each element a symbol. Most symbols are the first one or two letters of their English names. They always have a capital letter to start and if they have a second letter it is a small letter. These symbols were a lot quicker and easier to write than Dalton's symbols.

Different languages have different names for the same element. For example *azote* is Portuguese for nitrogen. However the element always has the same symbol.

Did You Know?

Some of the symbols like Pb for lead are from the Latin name of the element, e.g. plumbum.

C Why is it helpful for scientists to use symbols when communicating their ideas?

Summary Questions

1 Match the start and endings of the following sentences:

a) Dmitri Mendeleev was	a unique symbol that can be found on the Periodic Table.
b) All the elements have	that scientists who speak different languages can easily swap ideas.
c) Most elements have a symbol	the inventor of the Periodic Table.
d) Symbols for elements mean	that is the first one or two letters of their English name.

2 Find the symbols for the following elements:
a) oxygen b) helium c) titanium

3 Find the name of the element for the following symbols:
a) Si b) I c) C

4 Find out where the symbols for the following elements have come from:
a) Na b) K c) Sn

KEY WORDS

symbol
Periodic Table

Compounds and their Elements

- What is a compound?
- How are compounds different from elements?

Did You Know?

Alchemists were the first chemists. Instead of thinking about the world and writing down ideas, they did experiments. They used the results from the experiments to write ideas. Their main focus was to make gold from anything that was cheaper – they did not succeed!

What is a compound?

Alchemists discovered a lot of what they thought were elements in the 1600s. As technology improved it was possible to break these materials into simpler substances. This proves that some of these materials were not elements but more than one element joined together – a **compound**.

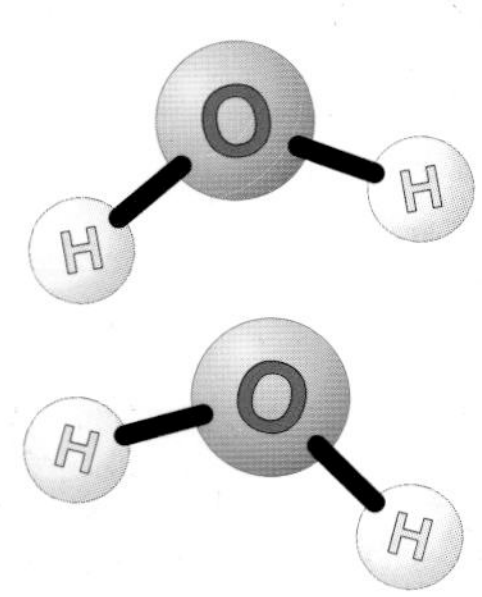

Particle diagram of water which is a compound

a Until the 19th century, only about 30 of the elements had been discovered. In the last 150 years, many more elements have been discovered. Can you think of any reason why?

A compound is a substance in which atoms of more than one type are chemically joined together. We can draw a particle diagram for a compound. There must be at least two types of atom joined (or bonded) together.

b Why is water called a compound?

A **molecule** is more than one atom chemically joined (bonded) together. Molecules can be made from just one type of atom (element) or more than one type of atom (compound). Oxygen forms 'two atom' (diatomic) molecules. Oxygen is in the Periodic Table and so it is an element. Hydrogen chloride forms 'two atom' molecules, but it is made from two different types of atom. This means that hydrogen chloride is a compound.

Stretch Yourself

Synthesis is a chemical reaction used to make a compound. The synthesis reaction to make water could be summarised by the following word equation:

hydrogen + oxygen ⟶ water

Can you find out the name and word equation for the synthesis reaction to make ammonia – an important chemical used to make fertiliser?

Ammonia is used to make fertilisers

activity

Making a compound

The properties of elements are often quite different from the compounds they make when joined together. For example, chlorine gas is toxic and was used in World War I to kill soldiers. Sodium metal will burst into yellow flames if water is dropped on it. However, when these two elements are chemically joined in a compound called sodium chloride, we eat it on our fish and chips! Your teacher might demonstrate this synthesis reaction.

mineral wool
mixture
red glow as the iron and sulfur react together
heat

You are going to compare the properties of iron, sulfur, a mixture of iron and sulfur and the compound iron sulfide.

- Observe the iron and sulfur with a magnifying glass and a magnet.
- Mix half a spatula of iron with half a spatula of sulfur. Observe the mixture with a magnifying glass and a magnet.
- Quarter-fill an ignition tube with your iron and sulfur mixture. Put a small plug of mineral wool loosely in the top of the tube. Using a test-tube holder, heat the mixture with a blue flame until it glows orange. Immediately take the ignition tube out of the Bunsen flame as soon as the mixture starts to glow and let it cool on a heat-proof mat.
- Your teacher will smash open the ignition tube. Observe the contents of iron sulfide using a magnifying glass and a magnet.
- Note all your observations in a table. How do the properties of the compound, mixture and its elements compare?
- Explain how you made your experiment safe.

Safety: Wear eye protection. Use in a well ventilated room.

Summary Questions

1. Copy the words below with their correct definition:

a) element	i) more than one type of atom, chemically joined
b) molecule of a compound	ii) more than one atom of the same element, chemically joined
c) molecule of an element	iii) all the atoms are the same
d) mixture	iv) more than one substance, not chemically joined

2. Carbon dioxide is a compound that makes up about 0.04% of the air:
   a) What key words could you use to describe this chemical?
   b) Draw a particle diagram of carbon dioxide as a gas.

3. Molecules made up of two atoms are called 'diatomic' molecules. Find three examples of elements that make diatomic molecules and three examples of compounds that make up diatomic molecules. What are molecules made of three atoms called? Can you find any examples?

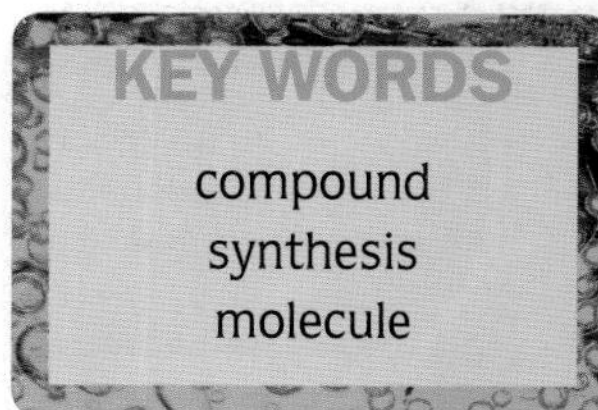

Compound Names

- How do I name a compound?
- How can I tell which elements are in a compound from its name?

magnesium + oxygen ⟶ magnesium oxide

Name that compound!

Many compounds are made between a **metal** and a **non-metal**. The metal name is always written first, with the non-metal name afterwards. When metals are in a compound their name stays the same. However, non-metals in a compound change the end of their name to **–ide**. When magnesium reacts with oxygen, magnesium oxide is formed. The magnesium is the metal and its name stays the same, but the oxygen changes to oxide.

The table shows how the names of some common non-metals change when they are in a compound with a metal:

Non-metal element	Name in a compound
Oxygen	oxide
Sulfur	sulfide
Chlorine	chloride
Bromine	bromide

a Predict the name of fluorine in a compound.

b Predict the name of the compound made between iron and sulfur.

c What elements would be in sodium oxide?

Sometimes a compound is made between three different elements, a metal, a non-metal and oxygen. This makes the ending of the non-metal change to **–ate**. Sodium sulfate would be made up of sodium, sulfur and oxygen.

Non-metal element	Name in a compound with oxygen
Nitrogen	nitrate
Sulfur	sulfate
Phosphorus	phosphate
Carbon	carbonate

d What elements are in calcium carbonate?

Did You Know?

When farmers put their fertilisers onto fields, some of the nitrates can be washed into streams and ponds by the rain. The nitrates cause the water plants to grow and this reduces the oxygen in the water and the fish can die. If the nitrates get into drinking water, it is believed to cause stomach cancer and blue-baby disease. Blue-baby disease is when a new-born baby's blood is starved of oxygen.

e What is the name of the compound made from magnesium, nitrogen and oxygen?

If a compound is made only of non-metals, the non-metal that is furthest to the left on the Periodic Table has its name unchanged. Hydrogen and chlorine are both non-metals, when they react together a compound called hydrogen chloride is made.

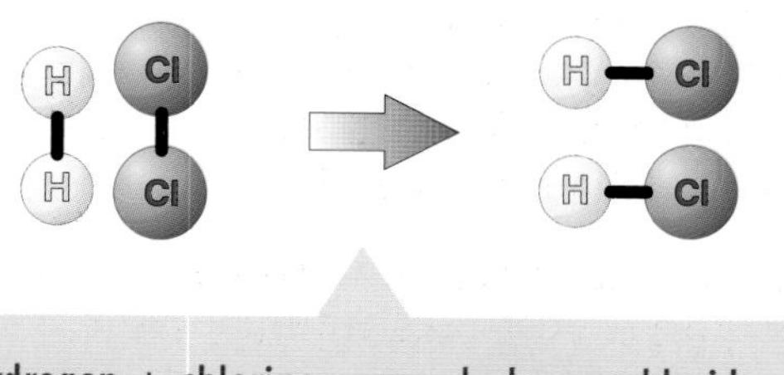

hydrogen + chlorine ⟶ hydrogen chloride

Stretch Yourself

Some compound names have prefixes before the parts of the name. 'Di-' means two and 'tri-' means three. This means that the scientific name for water is 'dihydrogen oxide', because it contains two hydrogens and one oxygen. Can you find other examples of compound names with prefixes?

hydrogen + oxygen ⟶ water

Summary Questions

1. Copy and complete the sentences below:

   Compounds can be formed between …

   The names of metals do not …

   The names of non-metals change …

   Compounds containing a metal, non-metal and oxygen …

2. Which elements are in the following compounds?
   a) silver oxide
   b) iron sulfate
   c) potassium chloride

3. Name the compounds made up of the following elements:
   a) lead and oxygen
   b) copper, carbon and oxygen
   c) sodium, iodine and oxygen

4. The chemical name for water is dihydrogen oxide. Some compounds can be named using the rules on this page, but others are still more commonly known by other names. Can you find some other examples of chemical and everyday names for the same compound?

KEY WORDS

metal
non-metal
-ide
-ate

Compound Formulas

- How do I write the formula of a compound?
- What does the formula of a compound tell me?

What's in the formula?

In C1.4, we learned that every element has its own symbol and we can find it in the Periodic Table. So, a compound can have a **formula** made up of the symbols of the elements that it contains.

a Which two elements are in sodium chloride, NaCl?

Did You Know?

The formula of a compound is always the same. H_2O is water, but H_2O_2 is hydrogen peroxide – used in hair bleach! One extra atom of oxygen makes all the difference.

The element that is closest to the left of the Periodic Table usually has its symbol written first. A small number to the right of a symbol means that there is more than one of that type of atom in the compound.

b What would be the formula of hydrogen bromide (which contains one of each type of atom)?

c How many atoms in total are shown in the formula of the compound calcium carbonate, $CaCO_3$?

We can use a model to help us predict the formula of a compound. Imagine each atom as a person and they want to hold hands, but how many hands does an atom have? The number of hands is the same as the combining power of the element. Look for the element in the Periodic Table. Then use the table below to help you work out how many hands the element has.

Group number	Number of hands (combining power)
1	1
2	2
3	3
4	4
5	3
6	2
7	1
0	0

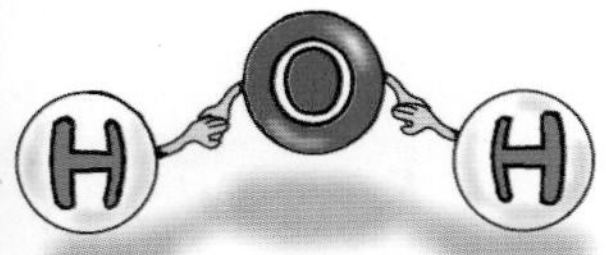

Hydrogen has one hand; oxygen is in Group 6 and has two hands. Water has the formula H_2O

Draw a diagram of your compound, making sure that you have all the elements at least once and every hand is being held. The hands are actually the **chemical bonds** and the number of bonds that an atom can make is called its **combining power**.

d What is the formula of calcium oxide?

activity

Splitting water

Water is a compound of hydrogen and oxygen. There are two atoms of hydrogen for every atom of oxygen. Electricity can be used to break down the molecules and make pure oxygen gas and pure hydrogen gas. The hydrogen gas can then be used in hydrogen fuel cells, which is a clean energy for cars.

Your teacher will run electricity through water in a special apparatus called a Hofmann voltameter. Look carefully at the diagram of the apparatus. Which of the tubes do you think the hydrogen will collect in? Why do you think this? How could you test the two gases? (Hint: Think back to Fusion 1.)

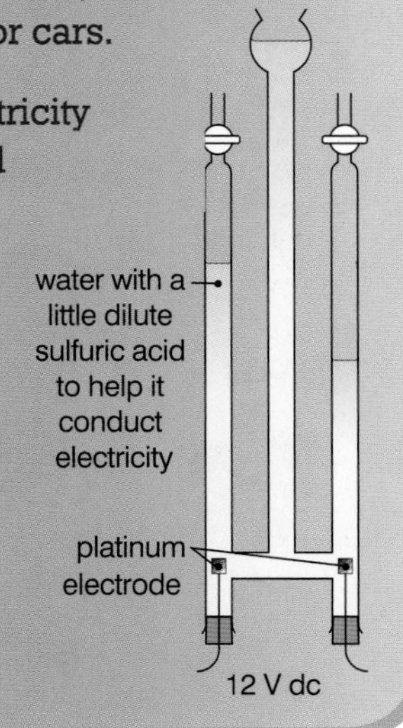

Stretch Yourself

Some compound formulas have brackets in them. The number outside the brackets is used to multiply all the elements in the bracket. For example, $Ca(OH)_2$ means one calcium atom, two hydrogen atoms and two oxygen atoms. Can you find other examples of compound formulas that contain brackets? How many of each type of element is in the compound?

Summary Questions

1. Copy and complete the passage using the words below:

   compound right chemically combining atom

   Compounds are made up of more than one element … joined. Each … can have a formula made from the symbols of the elements that it contains. The subscript numbers on the … of a symbol tell you how many of that … you have. The formulas can be predicted using the … power of the element.

2. Which elements are in the following compounds and how many atoms of each element are there?
   a) hydrogen peroxide, H_2O_2
   b) ammonia, NH_3
   c) sulfuric acid, H_2SO_4

3. Use combining powers to work out the formulas of the following compounds:
   a) lithium iodide
   b) calcium fluoride
   c) sodium oxide

4. Sulfate is an example of a molecular ion and has the formula SO_4^{2-} (an ion is a charged particle). The superscript number tells us how many hands the particle has or its combining power. Sulfate has two hands. Can you find out other examples of molecular ions and predict their combining power?

KEY WORDS

formula
chemical bond
combining power

Properties of Compounds

- How can I make a compound?
- Is a compound different depending on how I have made it?

Making compounds

Compounds can be made directly from their elements. However they can also be made by chemical reactions between other compounds.

a In the chemical reaction to make iron sulfide, what two elements would you start with?

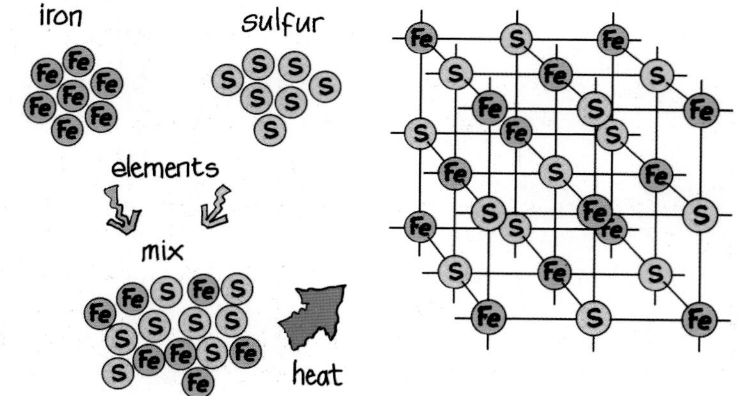

Next time you...

... drink a glass of water think about how many ways this water could have been made and reached you. In some parts of the country, the water you drink has been processed by nine people and has been purified at the sewage plant before it gets to you.

b Think back to Fusion 1. Can you explain two different chemical reactions that you could use to make sodium chloride (salt)?

Making carbon dioxide

Carbon dioxide is an important gas with many uses. Industry can collect the gas when power stations burn fossil fuels; and can be used to make cola fizzy!

You are going to make and test carbon dioxide using two different chemical reactions. How can you test for this gas? Do you predict that the carbon dioxide will have different properties if it is made in a different way?

It does not matter how you make a compound, its properties are always the same. This is because there is a set number and type of atom arranged in one way to form the compound.

activity

Method 1: Reaction of calcium carbonate with hydrochloric acid

Can you write a word equation for this reaction? Hint: A metal salt, a gas and water are the products of this reaction.

- Measure 3 cm^3 of hydrochloric acid into a test tube.
- Add half a spatula of calcium carbonate.
- Quickly put in the bung with the delivery tube going into a test tube of limewater.
- Make and record your observations.

hydrochloric acid and calcium carbonate

limewater

Safety: Wear eye protection. Limewater is an Irritant.

Method 2: **Thermal decomposition** of calcium carbonate

'**Thermal decomposition**' means using heat to break down a substance.

- Set up apparatus as shown.
- Heat on the blue Bunsen flame.
- Make and record your observations.
- How did you make your experiment safe?

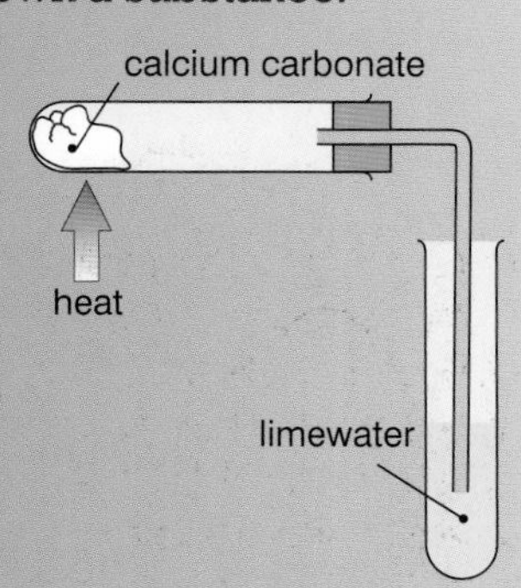

Safety: Wear eye protection and never touch hot glassware with your hands. Make sure you remove the end of the delivery tube from the limewater before you stop heating.

Summary Questions

1. Match each word with the correct definition and word equation:

a) chemical reaction	using heat to break down a chemical into simpler substances	magnesium carbonate $\longrightarrow$ magnesium oxide + carbon dioxide
b) thermal decomposition	a new substance is made	calcium + oxygen $\longrightarrow$ calcium oxide
c) synthesis	a chemical reaction used to make a compound	sodium + water $\longrightarrow$ sodium hydroxide + hydrogen

2. Hydrogen peroxide has the formula H_2O_2 and water has the formula H_2O.
   a) List the similarities and differences of these two compounds.
   b) Explain why these two compounds have different properties.

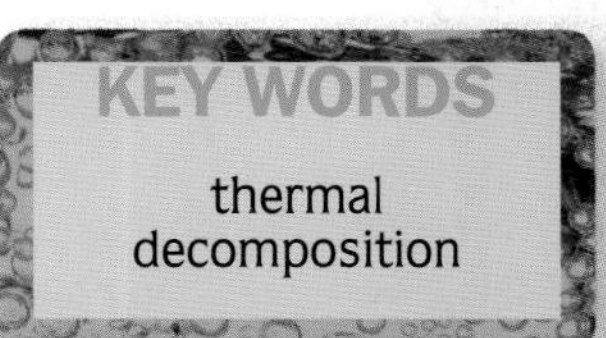

Word Equations

- How can I describe a chemical reaction?
- What is a word equation?
- How can I make a risk assessment

Link up to...

MATHS AND ENGLISH

Aim to write your word equations as you would an equation in maths (with an arrow instead of the equals sign); rather than a sentence as in English.

Write it all down

So, how can we describe on paper all these reactions that are happening? We can record chemical reactions in a many different ways. For example, we can use diagrams to show each atom in the reaction.

A diagram to show a chemical reaction to make water

A more scientific way to show chemical reactions is using **word equations**. Word equations show the chemicals we start with, called **reactants** and the substances we end up with, called **products**. We use an **arrow** (⟶) to show that the reactants go to the products. Never use an = sign as reactants do not equal the products.

silver nitrate + copper ⟶ copper nitrate + silver

(silver nitrate + copper: reactants; copper nitrate + silver: products)

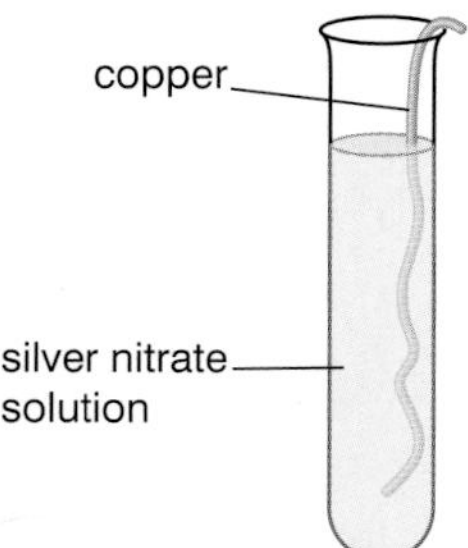

If you have big handwriting, make sure that the reactants are always written on the left of the arrow and the products on the right of the arrow instead of just carrying onto the next line.

If other chemicals or special conditions like a certain temperature are needed, these can also be written in a word equation, above the arrow.

Photosynthesis is the chemical reaction which happens in the chloroplasts of plant cells (see pages 32–35). For this chemical reaction to happen there has to be sunlight and the green chemical chlorophyll. This can be shown as a word equation:

$$\text{carbon dioxide + water} \xrightarrow[\text{chlorophyll}]{\text{sunlight}} \text{glucose + oxygen}$$

a What are the reactants and products in photosynthesis?

b Why are sunlight and chlorophyll written on the arrow in the word equation for photosynthesis?

activity

Describing chemical reactions

In real life it is important that scientists and people who work with chemicals complete risk assessments to minimise the hazards and the chance of an accident. Word equations can record all the chemicals and any special conditions and help in writing a risk assessment.

You are going to complete a number of different chemical reactions. Before you complete the reaction, you should write a word equation for each experiment and a brief risk assessment. Your risk assessment should focus on the chemicals that you are going to use. List any hazardous chemicals, their hazard symbol, how to reduce the risk and what to do if an accident happens. Think carefully about the presentation of the risk assessment as you will need to be able to refer to it quickly if there is an emergency. Check with your teacher before starting your experiment.

Station 1: Combustion

Burn a piece of magnesium ribbon in oxygen to make magnesium oxide.

Station 2: Neutralisation

React nitric acid and sodium hydroxide to make sodium nitrate and water.

Station 3: Thermal decomposition

Strongly heat copper carbonate to make copper oxide and carbon dioxide.

Record your observations for each chemical reaction in a table.

Did You Know?

There is a limited number of chemicals that we can use in schools. They are listed on Hazcards, which are written by CLEAPSS. Each chemical has its own card that tells you about the hazard and what to do in an emergency on the front side. On the reverse, there is a list of experiments that you can do with the chemical and what age group of pupil can do them. Teachers use these to write risk assessments for every experiment that you do in school.

Summary Questions

1. Copy and complete the following sentences:
   a) Reactants are the …
   b) Products are the …
   c) Word equations can be …
   d) An arrow is used in a …
2. General equations can be used to show the groups of chemicals that react and the groups of chemicals that they make. Complete some research to find the general equations for complete and incomplete combustion of a hydrocarbon.

KEY WORDS

word equation
reactants
products
arrow

Symbol Equations

- What is a symbol equation?
- How can I balance a symbol equation?

In the balance

In a chemical reaction no atoms are made and no atoms are destroyed. The atoms in the reactants are rearranged to make the products. This means that the mass of the reactants equals the mass of the products. This is because no matter has been made or lost. This idea will be explored in more detail in Fusion 3.

Did You Know?

In some chemical reactions the mass seems to change. If you take the mass of a nail and leave it on a window sill until it has rusted and then take the mass again, it will have increased. This is because the iron in the nail has gained oxygen from the air. This is known as an open system. However, if you did the same experiment in a sealed test tube, the mass of the tube and its contents would be the same at the start and end, as nothing extra can get in and nothing can get out. This is known as a closed system.

Magnesium metal reacting with hydrochloric acid

a What observations can you make that tell you the change when magnesium is added to hydrochloric acid is a chemical reaction?

Magnesium metal will react with hydrochloric acid. This can be shown in a word equation:

magnesium + hydrochloric acid ⟶ magnesium chloride + hydrogen

We can write the symbols and formulas for each of the chemicals:

$$Mg + HCl \longrightarrow MgCl_2 + H_2$$

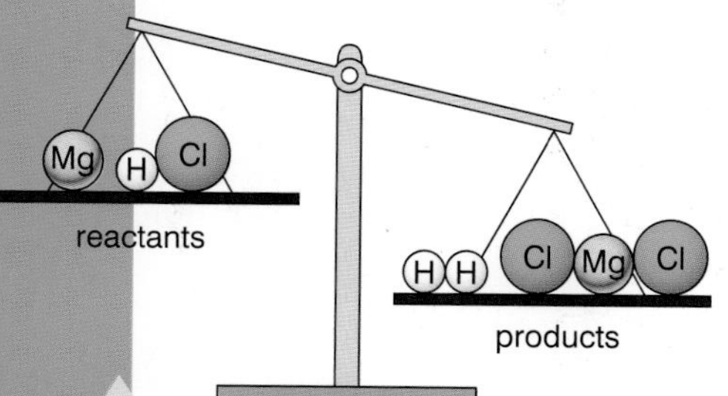

Which side is heaviest, the reactants or products?

There are more atoms on the products' side than on the reactants' side. We need to add extra chlorine and hydrogen, but we cannot add any small number (sub-scripts) to the formulas as this changes the compound. All we can do is add a large number in front of the formula or symbol.

The **balanced symbol equation** for this reaction is:

$$Mg + 2HCl \longrightarrow MgCl_2 + H_2$$

The balanced symbol equation is more useful than a word equation. The symbol equation tells us the quantities of the chemicals that are needed and produced in a reaction. Also we can track what happens to one type of atom.

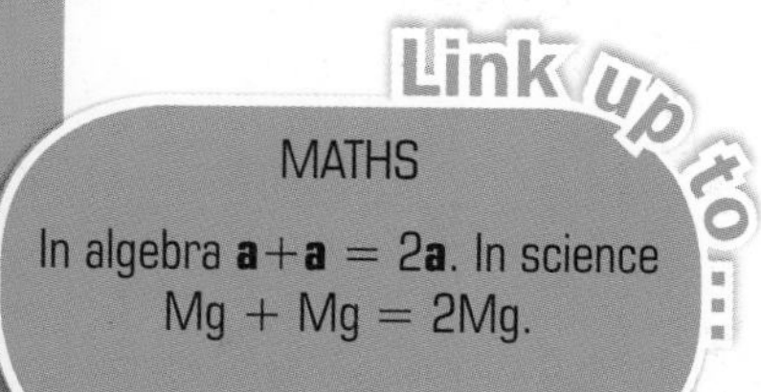

Link up to...

MATHS

In algebra **a**+**a** = 2**a**. In science Mg + Mg = 2Mg.

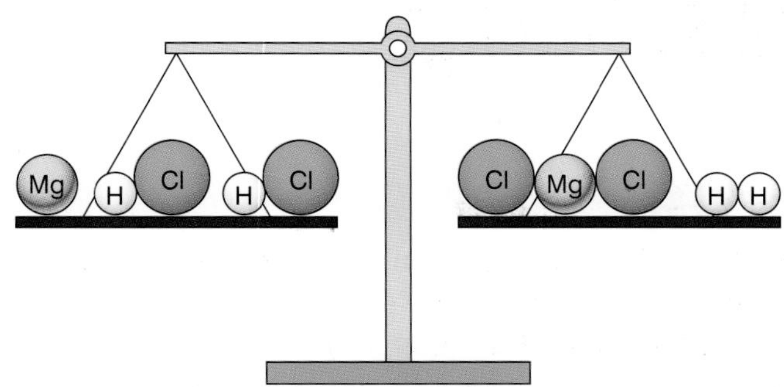

Imagine a set of scales; the reactants and products are balanced

b In terms of atoms and molecules, what do the following mean?

i) 3HCl ii) 2Fe iii) He

c Balance the following equation:

$$Mg + O_2 \longrightarrow MgO$$

Stretch Yourself

State symbols can be added to equations to show you what state the chemical is in. (s) means solid; (l) means liquid; (g) means gas and (aq) means dissolved in water. Some chemical solutions do not have water as the solvent. What do you think this might mean?

I_2 (ethanol)

Summary Questions

1 Match the start and endings of the following sentences:

a) All symbol equations	created or destroyed in a chemical reaction.
b) No atoms are	contain more information than word equations.
c) Symbol equations	must be balanced.
d) When balancing a symbol equation	you can only add large numbers in front of the formula or symbol.

2 Write balanced symbol equations for the following reactions:

a) carbon (C) + oxygen (O_2) ⟶ carbon dioxide (CO_2)

b) sodium (Na) + oxygen (O_2) ⟶ sodium oxide (Na_2O)

c) hydrochloric acid (HCl) + calcium (Ca) ⟶ calcium chloride ($CaCl_2$) + hydrogen (H_2)

3 Design a flow chart to help you remember how to write a balanced symbol equation.

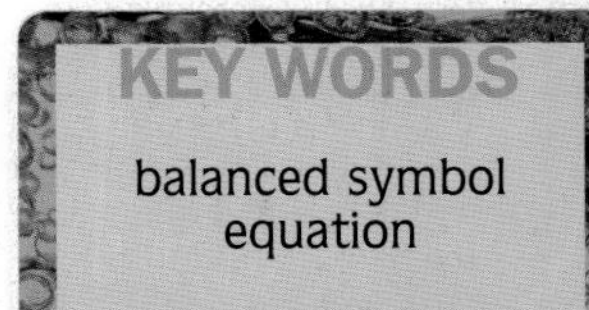

Mixtures

- What is the difference between elements and mixtures?
- What is the difference between compounds and mixtures?

Mixing it up

We can classify all materials as elements, compounds or mixtures. Elements and compounds are pure – they contain only one type of substance. Mixtures are made up of more than one type of substance, not chemically joined.

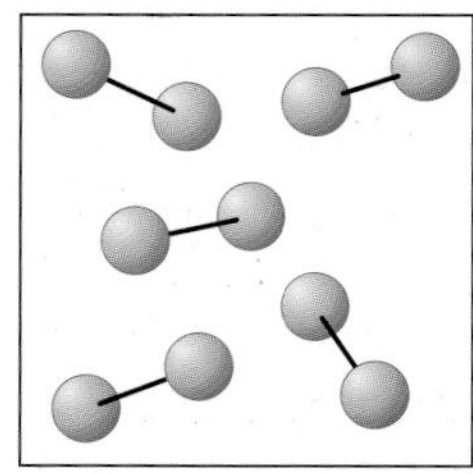

Element

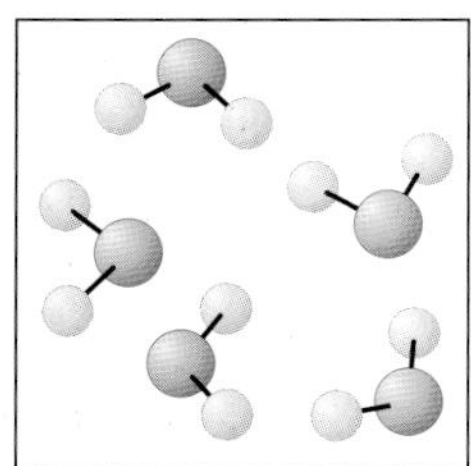

Compound

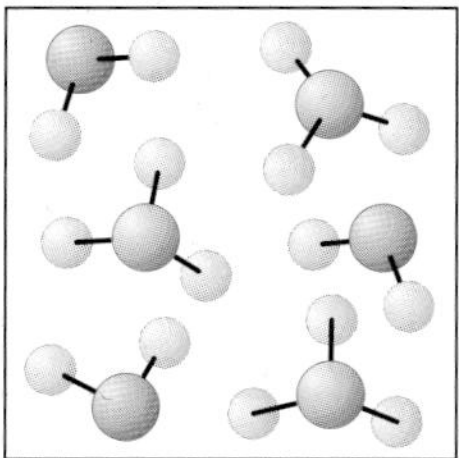

Mixture

Pure substances have a fixed **melting** and **boiling point**. They are unique to that pure chemical – like a fingerprint. There are books and databases that list the melting and boiling points of all known substances. Mixtures have a range of melting and boiling points. This is because each substance in the mixture changes state at a different temperature.

Did You Know?

Melting point is the temperature that a solid turns to a liquid or a liquid to a solid. It is the same as the freezing point.

a What are the melting point and boiling point of pure water?

b Why is seawater still a liquid when the temperature drops to $^{-}1$°C?

science@work

Police can use melting point techniques in crime labs. Special melting point apparatus can be used to find the melting point of an unknown solid quickly. A tiny amount of the substance is put into a very thin glass straw. This is lowered into the machine, where it is heated up. The tube is behind a magnifying glass. When the operator sees that the substance is melting they record the temperature. They can then use a data base to identify the substance, which could link a criminal to a crime scene.

A melting point apparatus

activity

Identifying compounds and mixtures

It is important to know the names of chemicals in a lab so that the hazards are known. Sometimes the labels fall off bottles and people are unsure what is in them. There are lots of different chemical tests to find out what the chemicals are. However, if you find the lost labels and the chemicals are relatively safe to heat, you can test their melting point and boiling point to identify them.

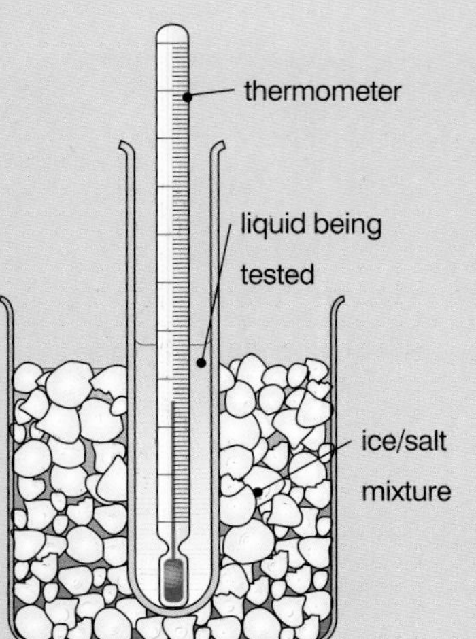

You will be given two different transparent liquids – one is pure water and one is tap water. Using melting and boiling points, design an experiment to prove which liquid is which. Write a step-by-step guide for your experiment and a risk assessment.

Check with your teacher before starting your experiment.

Summary Questions

1. Copy and complete the following sentences:
   a) The melting point of a pure ....
   b) Elements and compounds are ....
   c) Mixtures will melt or boil ....
   d) Mixtures are ....

   Copy and complete the table with facts about elements, compounds and mixtures.

   Hint: Your diagram should include particles; you should classify the melting and boiling points as being fixed or having a range.

Fact	Element	Compound	Mixture
Definition			
Diagram			
Melting point			
Boiling point			

2. Find out why salt is added to icy roads.

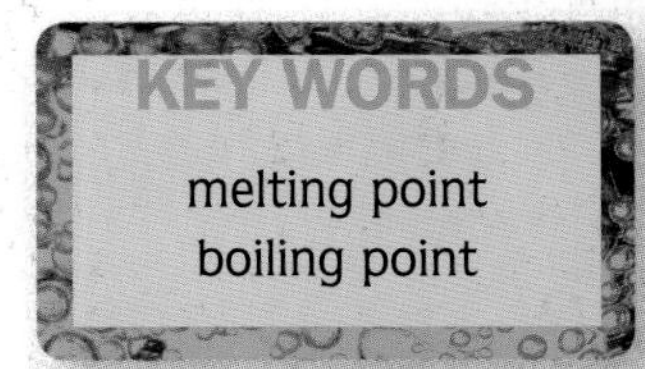

KEY WORDS

melting point
boiling point

Discovering Oxygen

- How do we know about oxygen?
- How have ideas of burning changed over time?

Oxygen is all around

Science is explained using theories. Theories are **ideas** that scientists have put together from results of experiments or by getting ideas from other theories. You cannot prove a **theory**; evidence can only support a theory. However, you can collect evidence that might prove a theory is wrong.

Oxidation is a type of chemical reaction where oxygen is added to something. Combustion, rusting and respiration are all examples of oxidation reactions. The connection between these reactions was first made in the 17th century.

Help Yourself

Look back at your earlier work and remind yourself about oxidation reactions. What reactant do they always have?

a Write a word equation for the combustion of charcoal, which is mainly carbon.

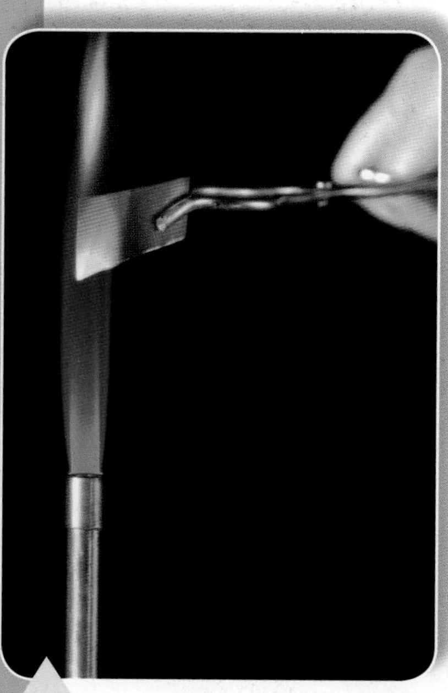

An example of an oxidation reaction

In the 17th century, two Germans, Johann Becher and Georg Stahl made a theory to explain combustion. They called it the **phlogiston** theory. They said that if a substance could be burnt it contained heat and a special substance called 'phlogiston'. Phlogiston could not be seen, touched or smelt. They used their theory to explain observations:

- Wood lost mass as it burnt because it lost phlogiston.
- A flame goes out in a sealed container because the air is too full of phlogiston.
- Charcoal is so full of phlogiston that it leaves very little ash when it has burnt.
- Metal ashes can be made back into metals if they are heated with charcoal, as the phlogiston moves from the charcoal to the metal ash.

Johann Becher (1625–1682)

Georg Stahl (1660–1734)

Did You Know?

The word 'phlogiston' comes from the Greek work *phlogios* meaning 'fiery'.

Phlogiston theory allowed scientists to group similar chemical reactions and explain them with just one theory. However, as science developed more measurements were made during experiments. In 1753, the Irish thinker Robert Boyle did experiments and endorsed phlogiston theory. The Russian, Mikhail Lomonosov said the theory was wrong as he proved that in a combustion reaction the mass did not change, as long as the container was sealed. Scientists then changed their theory to say that some phlogiston had negative weight and some had positive weight and that this explained Lomonosov's observations.

Antoine and Marie-Ann Lavoisier

Antoine Lavoisier, a French nobleman, made a balance that could measure to 0.0005 g and completed experiments on combustion with his wife Marie-Anne. From experiments, they developed two new theories: conservation of mass (you will learn more about this in Fusion 3) and an explanation of combustion. Lavoisier met Joseph Priestley to discuss his findings and then went on to try to claim credit for Priestley's work. Lavoisier experimented on air, found out that it was a mixture of gases and discovered **oxygen**.

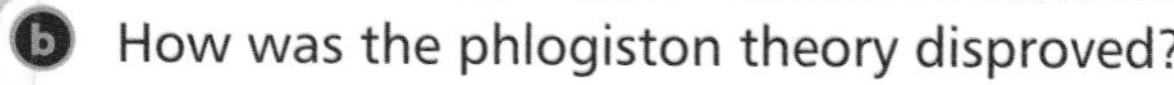

b How was the phlogiston theory disproved?

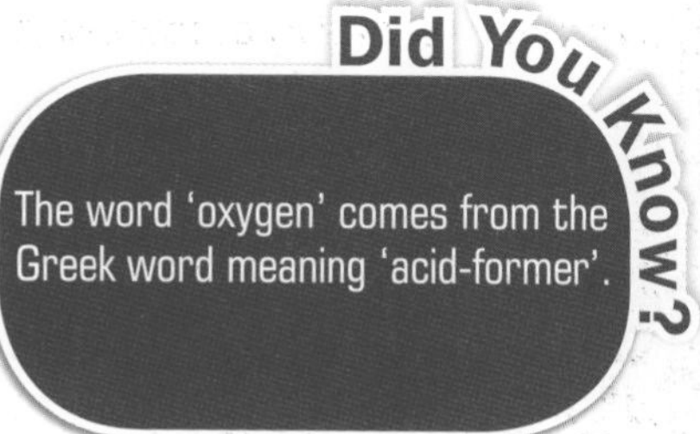

The word 'oxygen' comes from the Greek word meaning 'acid-former'.

Carl Wilhelm Scheele was a Swedish Chemist and the first to discover oxygen in 1773; a British philosopher, Joseph Priestley discovered oxygen by himself in 1774. However, Lavoisier is in the history books as the person who discovered oxygen, even though he did not find the element until 1775. This is because Lavoisier was the first to publish his work.

c How do you think these scientists communicated their ideas?

Summary Questions

1. Match each word with the correct definition:

a) theory	a vital part of a disproved theory used to explain combustion
b) idea	something used to explain the world around us, not using experiment
c) oxygen	an idea backed up with results from experiments and other ideas
d) phlogiston	an element that Lavoisier is often credited with discovering

2. Make a timeline to show how the phlogiston theory was made and dropped out of fashion. Include information about who discovered oxygen. You may want to go on the Internet and print-off pictures of the important scientists and even diagrams of their experiments to include in your timeline.

3. Choose an element and make a timeline to explain how it was discovered.

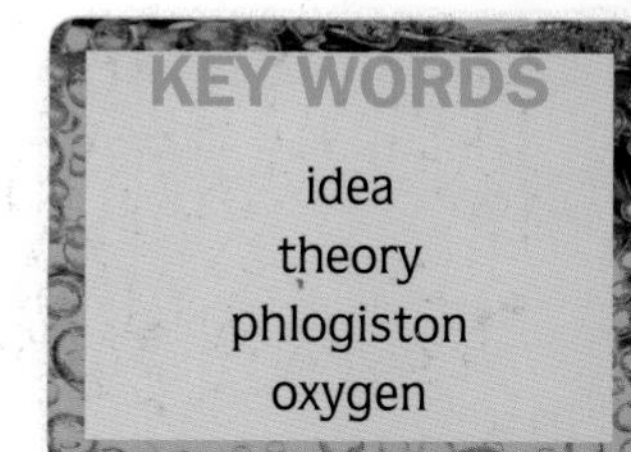

KEY WORDS

idea
theory
phlogiston
oxygen

Elements and Compounds – End of Topic Questions

know your stuff

▼ Question 1 (level 4)

Air is a mixture of different gases.

a Which element from the air do we use when we breathe in? Choose from the list below.

sulfur dioxide   oxygen   carbon dioxide [1]

b Which compound do we breathe out into the air? Choose from the list below.

sulfur dioxide   oxygen   carbon dioxide [1]

c Match each key word to the correct meaning.

(i) atom	only one type of atom and listed on the Periodic Table
(ii) element	more than one type of atom chemically joined
(iii) compound	smallest particle of a substance that can exist on its own

[3]

▼ Question 2 (level 5)

Most materials can be classified as metals or non-metals.

a Copy the following table of materials. Put *three* ticks in your table to show if they are metals or non-metals. The first one has been done for you.

Material	Metal	Non-metal
Steel	✔	
Sulfur		
Wood		
Cobalt		

[3]

b Look carefully at the table. Give an example of a magnetic *element*. [1]

c Why is steel used for making bridges? [1]

▼ Question 3 (level 6)

a What is an element? [2]

Below is a simplified part of the Periodic Table.

D

A   C   E

B

b Which letters represent metals? [2]

c Which letter is in Group 0? [1]

d Which letter is a gas? [1]

▼ Question 4 (level 7)

Iron can react with oxygen (O_2) in damp air and rust. Rust is iron oxide (Fe_2O_3).

a What sort of change has happened to iron when rust has been made? [1]

b How many atoms are there in the formula for iron oxide (Fe_2O_3)? [1]

c Write a balanced symbol equation for the reaction between iron and oxygen to make iron oxide (Fe_2O_3). [3]

How Science Works

Question 1 (level 4)

Dave and Bill are going to classify three different elements as metals or non-metals.

a Draw a table for Dave and Bill to record their results. [2]

Dave knows that metals often have a high density. To calculate the density of a substance you need to record the mass and volume.

b What is the unit of mass? [1]

c Explain how you could work out the volume of a sample of a material using water and a measuring cylinder. [3]

d Why might it be dangerous to measure the volume of some elements using the method in part (c)? [1]

Question 2 (level 5)

Maria decided to investigate how the percentage of oxygen affected how quickly magnesium burned. She decided to time how long the reaction took from start to finish.

a What is the symbol for the metal in this reaction? [1]

b What is the name of the compound made in this reaction? [1]

c On the packet of magnesium there was the following hazard symbol:

What does this hazard symbol mean? [1]

d On the oxygen cylinder there was the following hazard symbol:

What does this hazard symbol mean? [1]

e What is the independent variable in Maria's experiment? (The independent variable is the one Maria chose to vary.) [1]

f How could Maria record her results? Select *two* options from the following list.

table pie chart line graph [2]

Question 3 (level 6)

Hydrogen peroxide (H_2O_2) decomposes into oxygen (O_2) and water (H_2O). Special chemicals called catalysts can speed up this chemical reaction.

Merendeep wanted to investigate how different amounts of manganese dioxide affected how quickly oxygen was made. She used the following equipment:

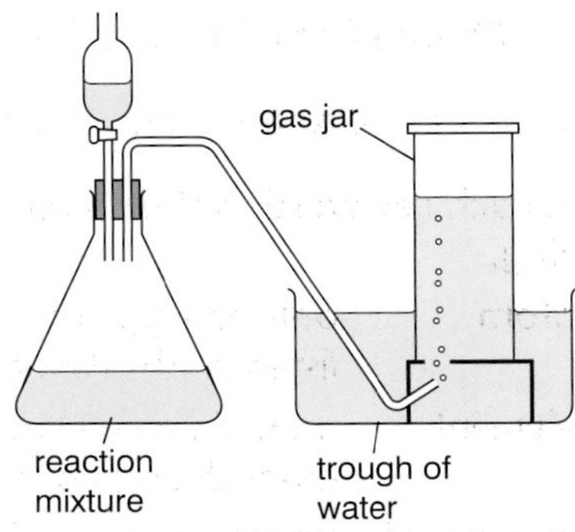

a Write a word equation for this reaction. [2]

b What piece of equipment could Merendeep have used to measure the volume of oxygen? [1]

Merendeep recorded her results in the following table:

Mass of manganese dioxide (g)	Volume of oxygen in 5 minutes (cm^3)
0	1
1	100
1.5	99
2	100
2.5	98
3	97

c What conclusions can you draw from Merendeep's experiment? [2]

d Why did Merendeep complete the experiment without any manganese dioxide? [1]

e Are Merendeep's results reliable? Choose from the list below.

Yes No Don't know

Explain your answer. [2]

Our Planet

Different rocks

- Rocks are very important building materials. Rocks, such as granite and sandstone, can be cut into bricks to make buildings.
- Other rocks, such as limestone, can take part in chemical reactions to make different building materials. We use limestone to help make glass, mortar and cement.

Buildings are made of rocks and products made from rocks

Did You Know?

Limestone, chalk and marble are all types of rock. They are made mostly of the same chemical compound – calcium carbonate. They have different properties and uses because they have different structures.

a Which rock can we use to make roof tiles?

Rocks are natural solids that are non-living. Most rocks are made of tiny grains that are stuck together. Rocks can be a mixture of different substances, which might be elements and/or compounds.

Did You Know?

Sand and clay are also classified as rocks!

b What does 'non-living' mean? (Hint: Think back to Biology.)

Different rocks have different properties. This makes them useful for different jobs. The properties of rocks can be explained by looking at their structure and what substances make the rock. Can you think of any rocks that you have used today?

Look at the two types of rock structure below:

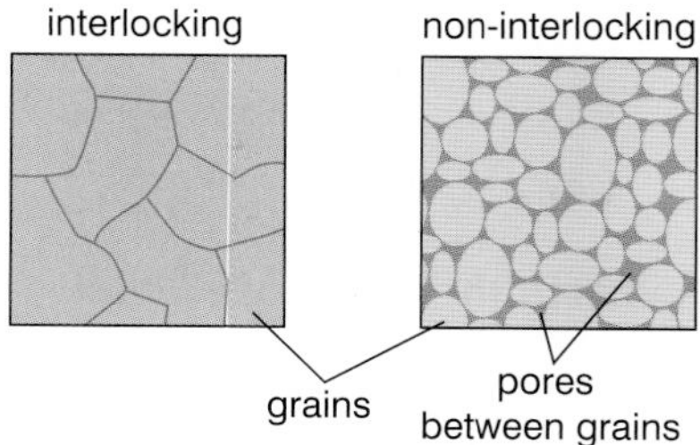

science@work

Geologists are scientists who study rocks and the Earth. They collect important information about our planet, so that we can begin to make models and better predictions about some natural disasters, such as earthquakes and volcanic eruptions.

c Which type of structure do you predict would be the harder rock?

'Rock hard'

The hardness of rocks is measured on a special scale called Moh's hardness scale. It has 10 numbers on it, 1 is as soft as talc and 10 is as hard as diamond. Find out how you can give a Moh's number to a rock.

activity

Studying rocks

Rocks are quarried or mined by companies to be sold for use in the building industry. By looking at the structure of rocks we can predict the properties and suggest how the rock was made. These are important pieces of information that companies need to have in order to decide if it is worth spending their money quarrying the stone.

- Collect a sample of one type of rock.
- Use a hand lens to examine the surface of the rock.
- Can you see the tiny grains that the rock is made from?
- What do they look like?
- Can you see the spaces between the tiny grains?
- Take a mounted needle and scratch the surface. Is the rock hard or soft?
- Repeat these steps for each type of rock.
- Record your results in a table like the one below:

Rock sample	Structure	Hardness

- Can you group or classify the rocks?
- Think about the properties of each rock that you have studied. Suggest some uses for each rock.

Igneous Rocks

C2.2

- What is igneous rock?
- How is igneous rock formed?
- How can I get reliable results?

Formation

Under the surface of the Earth there is molten rock called **magma**. Sometimes the magma gets onto the surface of the Earth from a volcano and this is called **lava**. When the magma or lava cools it solidifies (freezes) and forms a solid rock. We call this **igneous** rock.

a What is lava?

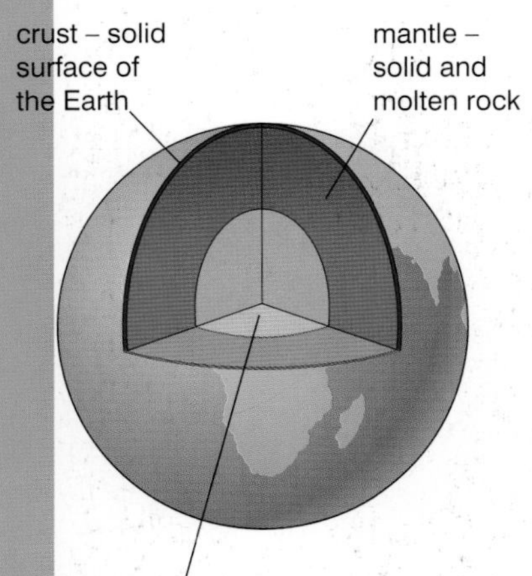

Structure of the Earth

activity

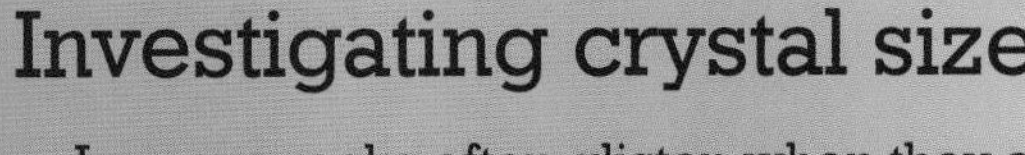

Investigating crystal size

Igneous rocks often glisten when they are held to the light. This is because they contain crystals. The size of the crystals depends on how long the lava or magma had to cool.

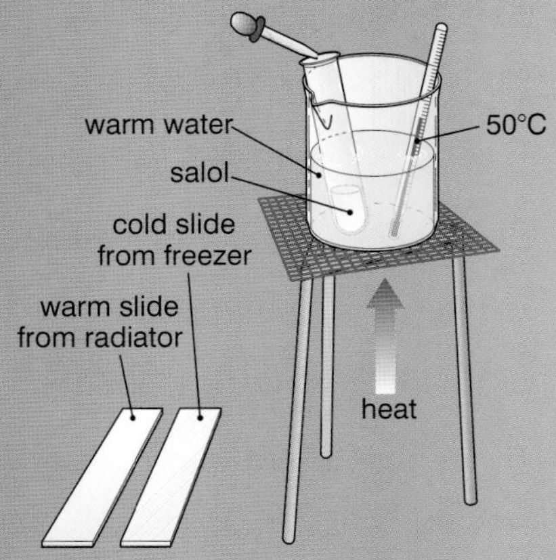

- Collect a warm microscope slide and a cooled microscope slide.
- Using a dropping pipette, put a small drop of molten salol onto each slide.
- Using a mounted needle, gently lower a cover slip onto the top of the salol.
- Observe the two slides.
- When the salol has solidified, have a look at the crystals under a microscope. Sketch a diagram of what you see.
- What have you found out about the size of crystals and rate of cooling?

Safety: Wear eye protection. Use a mineral wool plug to stop vapour escaping.

Did You Know?

Basalt is an example of igneous rock. Most of the ocean floors are made of basalt.

The results of an experiment are more likely to be **reliable** if every time you repeat the experiment you get a similar result. However, you might be repeating the same mistakes each time. If other people do the same experiment and get the same results, then the results are even more reliable.

The results are **accurate**, if the results are the very close to the true values. One way to find the true values is by looking at results that have been published in reliable sources like books.

Properties

Igneous rocks often have an interlocking structure. This makes igneous rocks non-porous and hard. They also often have crystals in them. The longer it takes the molten rock to cool down, the bigger the crystals it will form. This is because it takes time for a crystal to grow.

Igneous rocks do not have fossils in them. This is because they are made from molten rock at very high temperatures. Any fossils in the rock would have been destroyed when the rock melted.

b Where might you find igneous rocks with big crystals?

Link up to...

HISTORY

Pompeii was an ancient Roman town in Italy. It was near the volcano called Mount Vesuvius. When the volcano erupted almost 2000 years ago it killed many people. The town was covered in lava and ash. This cooled and made igneous rock which has saved a lot of history, including preserving the shapes of human bodies encased in ash. As the volcano erupted, bubbles of gas were trapped in some of the magma and made a rock called pumice (you may have used pumice to rub rough skin off your feet). Many people were killed as pumice hit them and the weight of the rock made buildings collapse.

Pompeii

Summary Questions

1 Match up the key words with their definitions:

a) magma	molten rock under the surface of the Earth
b) lava	rock made from solidified magma or lava
c) igneous rock	results that are very similar if the same experiment is repeated by other people
d) reliable	results that are very close to the true values
e) accurate	molten rock on the surface of the Earth

2 Make a list of characteristics that you would look out for that would tell you a sample was an igneous rock.

3 Complete some research to find out three different examples of igneous rocks and a use for each.

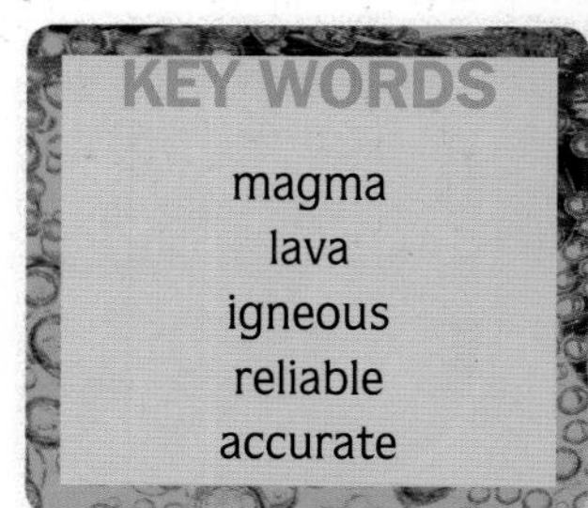

KEY WORDS

magma
lava
igneous
reliable
accurate

Sedimentary Rocks

- What is sedimentary rock?
- How is sedimentary rock made?
- How can I evaluate my investigation?

How are they formed?

Sedimentary rocks, such as limestone, are mainly made under water in our seas, rivers and oceans. Small pieces of rock, called sediment, get carried by moving water. The sediment is made from other rocks that have been broken down. Eventually they are dropped (**deposited**) at the bottom of the water.

These small pieces make the grains in the rock. They settle in layers and get squashed together (**compaction**). Minerals dissolved in the water then stick the small rock pieces together (**cementation**) and this forms the sedimentary rocks.

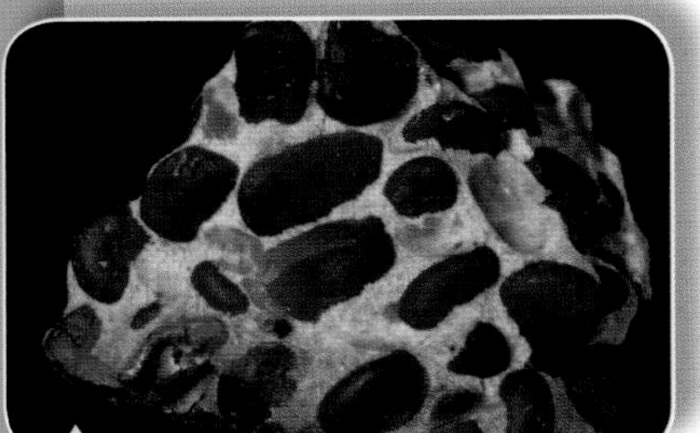

Conglomerate

Sandstone

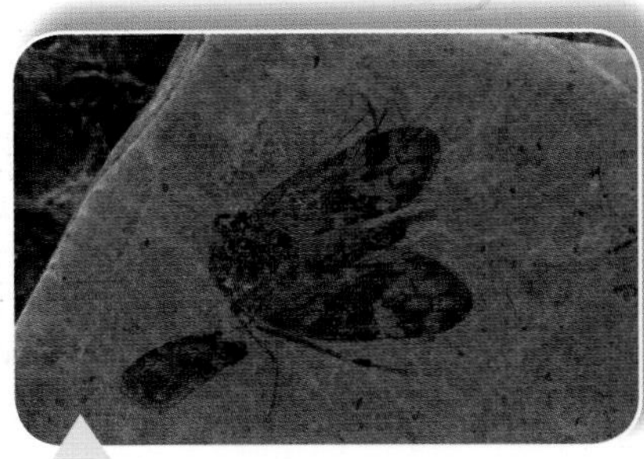

Mudstone

a Name some sedimentary rocks.

Beds of sedimentary rock

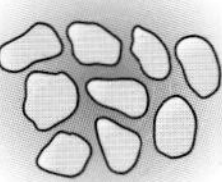

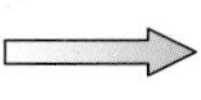

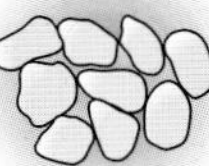

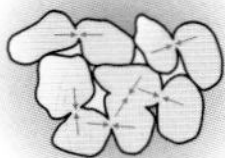

Sand grains as they are deposited; water fills spaces between grains.

As more sediment build up, the pressure increases and pushes grains of sand closer together, squeezing out the water.

The edges of the grains can stick together under this pressure, forming solid rock.

Sedimentary rocks form layers known as beds. Usually, the deeper the bed the older the rock is. Layers of sedimentary rocks can also trap dead plants and animals. Over millions of years, these organisms turn into fossils.

The rocks have a non-interlocking structure and so they have small gaps between the grains. These gaps let water get into the rock and this rock is described as porous. Sedimentary rocks tend to be softer and more crumbly than other rocks.

A fossil

Did You Know?

Fossils are the shape of the plant or animal, but the living material has rotted away and some parts of the organism calcify (becomes a calcium compound). This means you do not know what colour the plant or animal was, but the pattern of the leaves or skin may still be seen.

activity

Investigating sedimentary rock

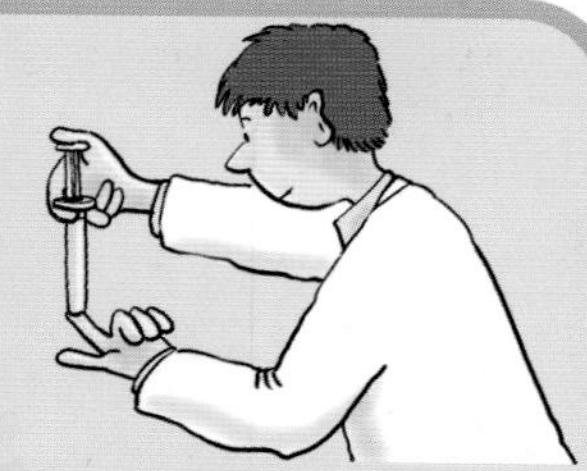

Sedimentary rocks take thousands of years to form. However they all have a similar appearance with layers and grains. Geologists can look at a sample of rock and identify if it is sedimentary rock by looking for these properties.

You are going to make a model of sedimentary rock using sand, clay and plaster of Paris. You will investigate which mixture of substances makes the strongest model of a sedimentary rock. Which material will model the grains? Which material will model the cement?

- Get a syringe, with the end cut off. Use petroleum jelly to grease the inside of a syringe and plunger.
- In a beaker mix some sand with water.
- Pack the wet sand into the syringe and put your finger over the end and squeeze out the extra water.
- Now push out the sand cylinder onto blotting paper and label as 'sand'.
- Using plastic spoons to measure the powders, make at least three different mixtures of sand, clay, plaster of Paris and water in different beakers.
- Make cylinders as before.
- Let the models dry.
- Put one model at a time between two heat-proof mats and add slotted masses on the top mat. Record in a table what mass is needed to break each mixture.
- What are the errors in your experiment? Are they random or systematic? Explain how you have classified the errors. How could you improve your investigation? (See below.)

Safety: Wear eye protection and wash hands after making the models.

In every experiment there are **errors**; this means that the results may be less reliable and less accurate. If the same error happens every time the experiment is done, we call this a 'systematic' error. This error will be the same in every result and so the conclusion is not changed, but the results are not accurate. If the error is sometimes there and other times not, then this is 'random' error and the results are unreliable and not accurate.

Summary Questions

1. Copy and complete the passage using the words below:

   beds sediment older grains dissolved

   Sedimentary rocks are made under water. …is deposited into layers, these are compacted (squashed) and minerals stick the different … together to make the rock. The minerals for cementation come from … chemicals in the water. The layers of sedimentary rocks are known as …. The deeper the bed is from the surface of the Earth, the … the sedimentary rock is.

2. Find out some uses of sedimentary rocks. Which properties make them suitable for their job?

KEY WORDS

sedimentary
deposited
compaction
cementation
error

Metamorphic Rocks

C2.4

- What is metamorphic rock?
- How is metamorphic rock made?
- How can I use a model to explain how rock forms?

All change

Metamorphic rocks are made from other rocks that have been changed by **heat** and/or **pressure**. If a volcano is formed near limestone, the magma will cause extreme heat. Limestone is then changed into marble. The structure changes and so the metamorphic rock will have different properties to the rock that it was made from. Sometimes the extreme heat and pressure can cause chemical reactions in the rocks as well.

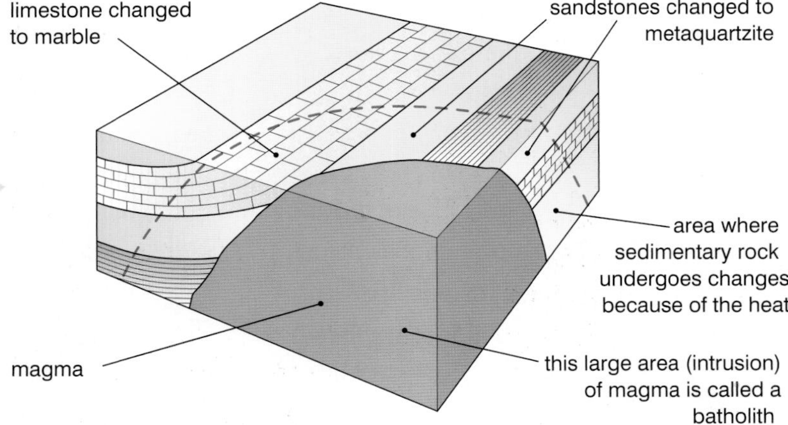

Limestone changes to marble

a Look carefully at the diagram above. What do you think would happen to the layers and fossils in the limestone as it is changed to marble?

When sedimentary rocks change into metamorphic rocks we say they have been metamorphosed. Sandstone is changed by heat and pressure to make metaquartzite.

'Morphic' means 'change'. Metamorphic rocks are formed mainly from sedimentary rocks which have had heat and/or pressure acting on them. Shale is a sedimentary rock that can be changed under heat to make hornfels. However, when heat and pressure is present, shale becomes slate. If slate is then heated and squashed more, schist is formed. If very high **temperatures** and pressures happen then schist can make gneiss.

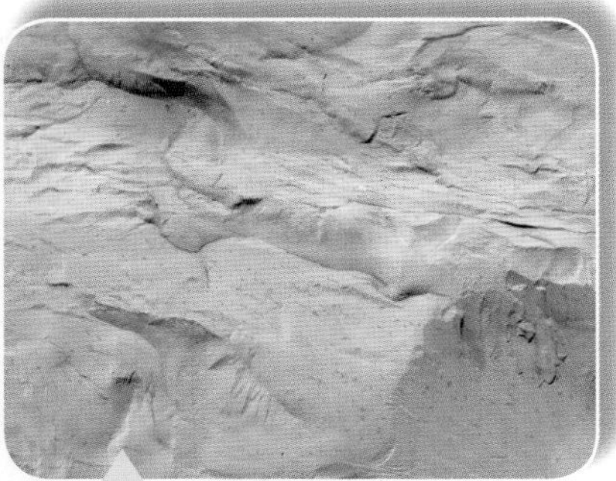

Limestone

Marble

b List two examples of sedimentary rocks and the names of the metamorphic rocks that they make.

c Draw a flow chart to show how shale can become gneiss.

Metamorphic rocks form layers and often have interlocking crystals. The structure of metamorphic rocks means that there are no spaces for water to get into the rock so they are non-porous. Metamorphic rocks may have fossils, but the heat and pressure change their shape or destroy them.

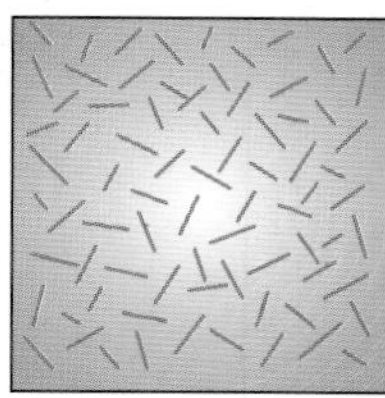

mudstone

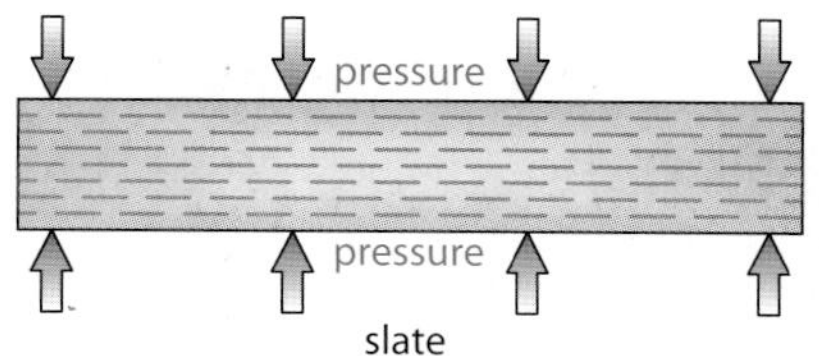

slate

Modelling metamorphic rock

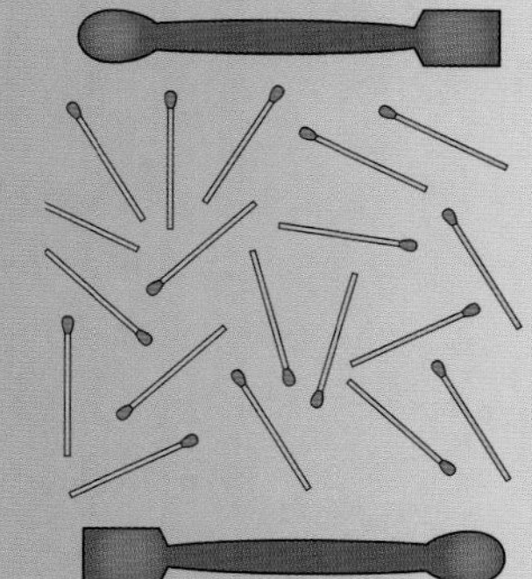

The properties of metamorphic rocks can be explained by their structure. Models can help scientists understand and explain observations. You are going to use matches and two spatulas to model how sedimentary rocks become metamorphic rocks.

- What do the matches represent?
- How will the matches be arranged when sedimentary rock is being modelled?
- What do the two spatulas represent?
- How will metamorphosis be modelled?
- How will the matches be arranged when metamorphic rock is being modelled?

Stretch Yourself

There are three grades of metamorphic rock. Low-grade metamorphic rock is formed from low pressures and/or heat on sedimentary rock. These rocks may still have fossils in them and no mineral veins. Medium-grade metamorphic rock is formed under higher temperatures and pressure. Medium-grade metamorphic rock has mineral veins that can be seen. What is high-grade metamorphic rock? How is it formed?

Summary Questions

1. Match the start and endings of the following sentences:

a) Metamorphic rocks are	an example of metamorphic rock made from limestone.
b) Marble is	made when other rocks undergo changes because of heat and/or pressure.
c) Metamorphic rock contains	crystals and layers.

2. Make a list of characteristics that you would look out for that would tell you a sample was a metamorphic rock.

3. Complete some of your own research to find out the names of metamorphic rocks and how they are made. Record your information in a table.

KEY WORDS

metamorphic
heat
pressure
temperature

Which Rock?

▸▸ How can I classify and identify rocks?

Classifying rocks

There are three types of rock:

- Igneous
- Sedimentary
- Metamorphic.

All the rocks on the Earth's surface can be put into one of these three groups. By carefully studying the rock and testing its properties it can be classified.

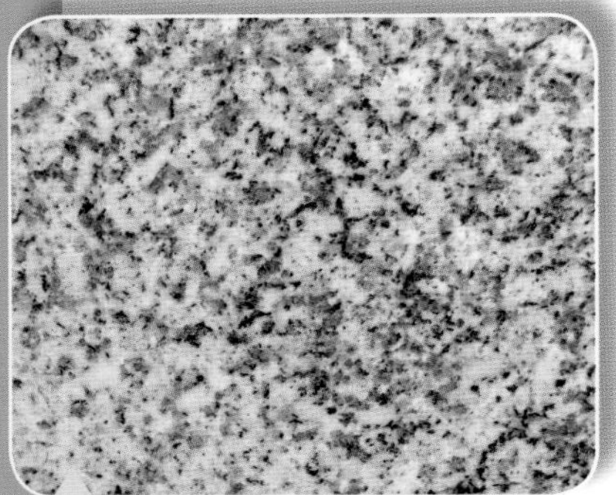

Igneous (granite)

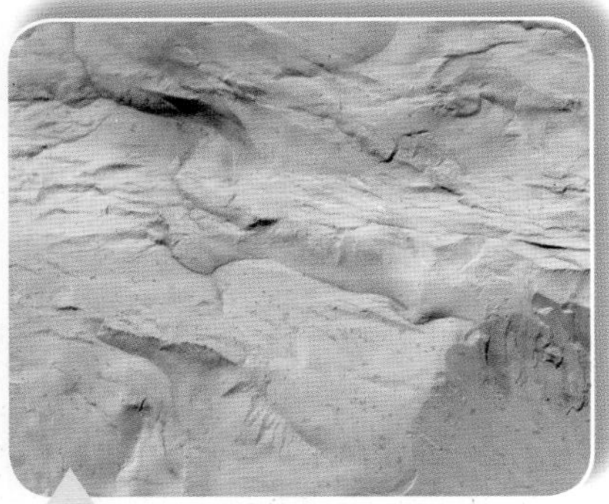

Sedimentary (limestone)

Metamorphic (slate)

a Make a list of properties that are the same and those that are different for sedimentary and metamorphic rocks.

b Make a list of properties that are the same and those that are different for igneous and metamorphic rocks.

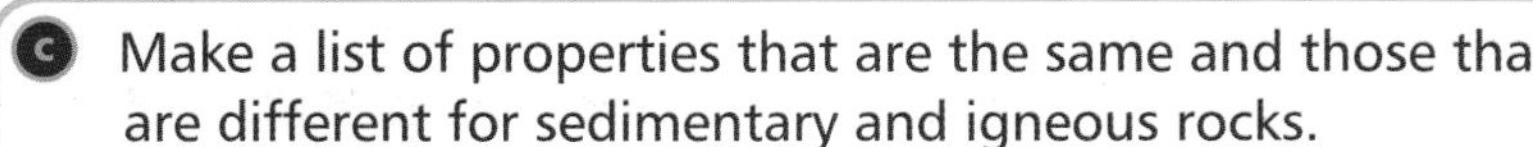

c Make a list of properties that are the same and those that are different for sedimentary and igneous rocks.

Each type of rock is made in a different way and its structure decides the general properties of that group of rocks.

Next time you...

... go shopping, have a good look around the town and see how many different types of rock you can spot. Maybe the roof tiles, the flooring, even a bottle of talc! Can you identify which family of rocks they belong to?

Keys that rock!

You will have used keys to help you classify things before. There are two types of **key**: **branching** or **dichotomous**. It is important that the questions in a key have definite answers that cannot lead someone to take the wrong path. Look at the examples on the next page.

Did You Know?

The youngest rocks on Earth are igneous rocks. We can use fossils trapped in sedimentary and metamorphic rocks to age the samples.

1. Does the shape have corners?

   Yes – Go to question 2.

   No – Circle.

2. Does the shape have 3 sides?

   Yes – Triangle.

   No – Go to question 3.

3. Does the shape have 4 equal sides at right-angles?

   Yes – Square.

   No – A polygon with 4 or more sides, not a square.

Dichotomous key to help you identify shapes

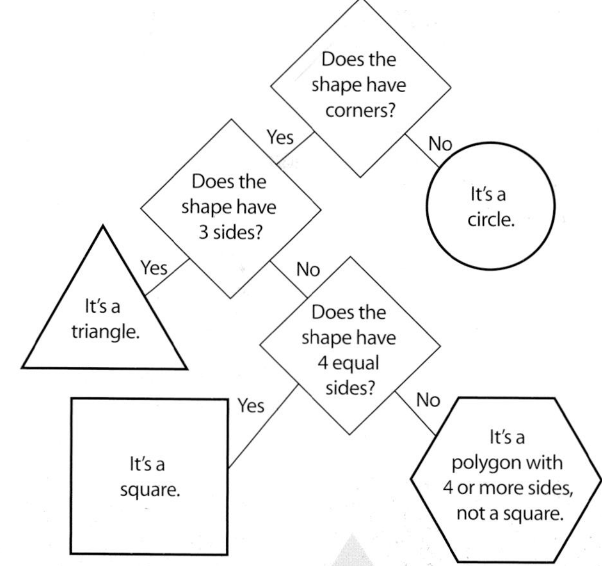

Branching key to help you identify shapes

activity

Creating a key to classify rocks

When geologists go out on field trips they may come across samples of rock they find difficult to identify. By using a key they can quickly decide which type of rock group it belongs to and give its name.

- Your teacher will give you some samples of rocks and their names.
- First you need to decide which rock group they belong to.
- Then make a key to help people to identify these rocks.
- How could you test your key? How could your key be improved?

Summary Questions

1. Copy and complete the sentences below:

   There are three groups of …

   Keys are used to …

   Keys can be described as …

   General properties of rocks can be explained by …

2. Explain why it is important to classify rocks as igneous, metamorphic or sedimentary.

3. Limestone, chalk, marble and a sea fossil are all mainly calcium carbonate. Make a key to help you identify each sample.

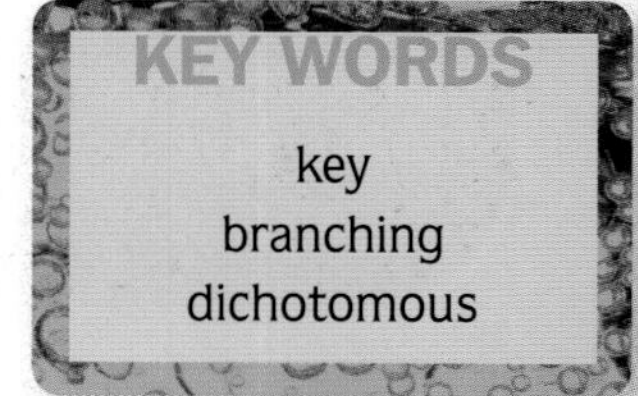

KEY WORDS

key
branching
dichotomous

Weathering

C2.6

- What is weathering?
- How can plants and animals change rocks?
- How can I plan an investigation?

Rocks on hills, mountains, cliffs and buildings are broken up by processes called **weathering** and erosion. There are three types of weathering:

- **biological**
- **chemical**
- **physical.**

If this doesn't weather soon I'm going home.

a What kinds of thing do you think cause each type of weathering?

Biological weathering is when plants, animals or microorganisms attack the rock. (You will have learnt about microbes in Fusion 1.) This can cause the rock to weaken and break into smaller pieces.

Roots of plants can grow into the surface of a rock and weaken it, so that it crumbles away. As a crack starts to be made, soil can get trapped and seeds can grow. This means that there are even more roots to break down the rock.

Did You Know?

Some microorganisms feed on rocks! Many metals are locked up in rocks. The rocks contain metal compounds known as minerals. A chemical reaction is needed to release the metal from its mineral. After we have dug up the mineral from the Earth, microorganisms can be used to 'eat' the mineral and release the metal. This is a lot cheaper and more environmentally friendly than using traditional chemical reactions.

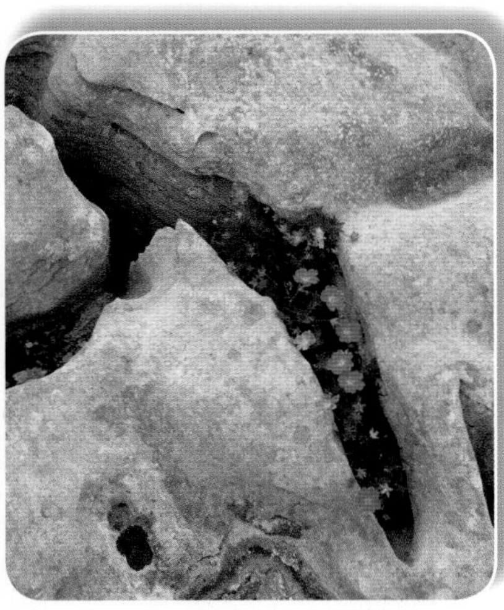

Some animals live under the ground in burrows. These animals can scratch at the surface of rocks and dig at the rocks, breaking them into small pieces.

b Give one example of a burrowing animal that could cause biological weathering.

activity

Biological weathering

When you are planning an experiment you need to think about what results you need. The independent **variable** is the variable that you want to change. The dependent variable is the variable that you measure to judge the effect of changing the independent variable. The control variables are other factors that could affect the results, so you need to choose values for them and stick to them.

When town planners are deciding where to build roads they have to think carefully about biological weathering. It is important to clear any roots that could damage a road surface. They may also need to re-locate protected animals like badgers and other burrowing animals that could damage the road. Town planners will choose what materials to make the road from and the depth of each layer.

You are going to design an experiment to find out the effect of oak trees on different road surfaces. You will investigate four different road surfaces: concrete, tarmac, cobbles and sand. You need to list the dependent, independent and control variables, write a brief method and draw a labelled diagram.

Summary Questions

1. Copy and complete the passage using the words below:

   rock physical microorganisms biological pieces

   Weathering is the breaking down of rock into smaller …. There are three types of weathering: chemical, … and biological. Plant roots grow into rocks, animals burrow into rocks and … can eat rocks. These are all examples of biological weathering. We call it … weathering when plants, animals or microorganisms attack the … and break it into smaller pieces.

2. Jeff is a keen gardener, but he has noticed that his concrete garden path keeps breaking up and has lots of cracks. What advice would you give to Jeff to help keep his path in good condition?

3. Use a digital camera to take some photographs of biological weathering happening to your school.

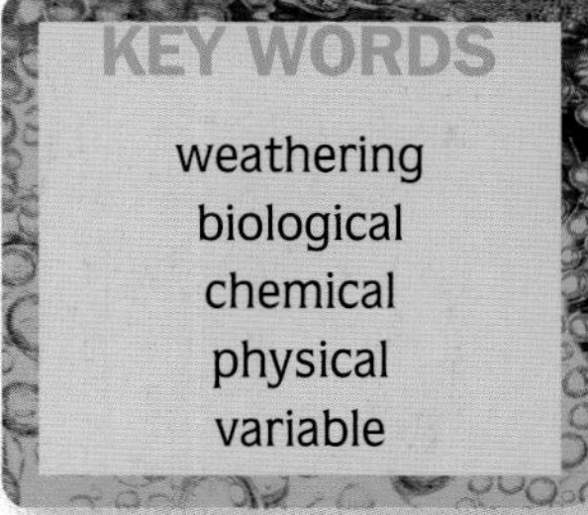

KEY WORDS

weathering
biological
chemical
physical
variable

Chemical Weathering

- What is chemical weathering?
- How does chemical weathering change rocks?

Write a word and balanced symbol equation for the formation of carbonic acid from water and carbon dioxide.

It's raining chemicals!

Rain water is actually a mixture of chemicals, including **carbonic acid** (H_2CO_3). This is made by carbon dioxide from the air dissolving in rain. It has a pH of 5.5 and is a weak acid. Some rocks contain chemicals that will react with the acid in rain water.

This is a **neutralisation** reaction (see Fusion 1) and this will break the rocks down. The breaking down of rocks by chemical reactions is called **chemical weathering**.

a What type of chemical is likely to react with rain water?

b What is the definition of a neutralisation reaction?

activity

Chemical weathering by acid

Caves and pot holes can be found in some rocks. The acid in rain water gets into cracks and chemically weathers the rock to make holes and tunnels. Some of the products of the chemical weathering will be dissolved in the rain water. As the water runs through the caves and pot holes it will evaporate and the dissolved chemicals are left behind making stalactites and stalagmites.

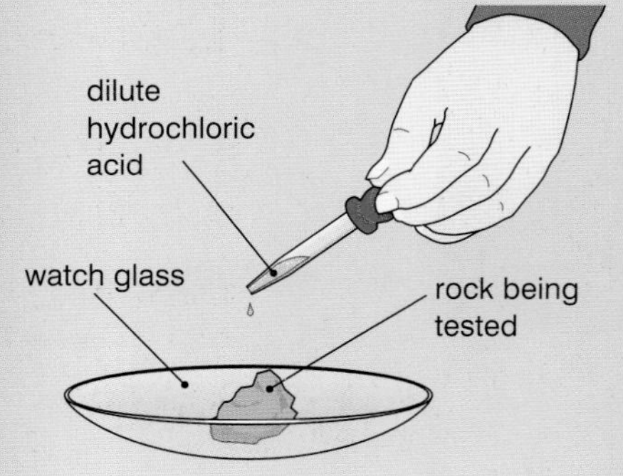

You are going to investigate the effect of acid on three different rocks – limestone, marble and chalk. You are going to study their reaction with hydrochloric acid. This is a stronger acid than carbonic acid and so the results of chemical weathering will happen a lot quicker.

- What is the main substance that makes up these three rocks? Make a prediction about what observations you will make and what this means.
- Put the piece of the rock in the centre of a watch glass.
- With a mounted needle and magnifying glass observe the surface of the rock.
- Using a dropping pipette add a few drops of acid onto the surface of the rock.
- Re-observe the surface of the rock and record your observations.

Safety: Wear eye protection and wash hands after setting up the experiment.

- Write a general word equation for the reaction between a metal carbonate and acid. Hint: A metal salt, water and a gas is made. How could you prove what the gas is?

Chemical weathering can also happen when other chemical reactions attack the rock. **Oxidation** reactions (where oxygen is added to a substance) can cause chemical weathering.

c What is the definition of an oxidation reaction?

activity

Chemical weathering – oxidation

Weathering is a slow process and is difficult to observe. We can use satellites to take photos of the Earth's surface over a long time and they can be compared to see the effects of weathering on the landscape.

You are going to investigate the effect of acid and oxidation on granite. You are going to study its reaction with hydrochloric acid (which will model acids in soils) and hydrogen peroxide (which releases oxygen, modelling oxidation). Make a prediction about what observations you will make and what this means.

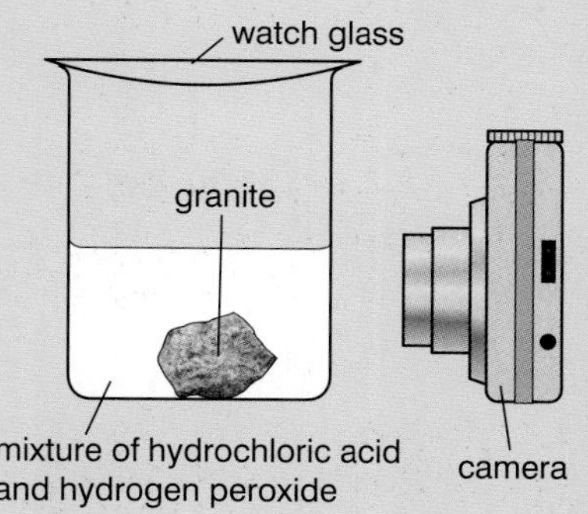

- Put the sample of rock in a beaker.
- Use a mounted needle and magnifying glass to observe the surface of the rock.
- Half-fill the beaker with a mixture of hydrochloric acid and hydrogen peroxide. Cover the top of the beaker and put it in a place where it will not be disturbed.
- Set up a digital camera to take a photo of the rock every few hours for a week.
- Re-observe the surface of the rock and the photos. Record your observations.

Safety: Wear eye protection and wash hands after setting up the experiment. If chemicals used are corrosive make sure eye protection is of the chemical splash proof variety.

Summary Questions

1. Match up the key words with their correct definitions:

a) chemical weathering	a weak acid found in rain water
b) oxidation	chemical reactions that break down rock into smaller pieces
c) carbonic acid	a chemical reaction that adds oxygen to a substance
d) neutralisation	a chemical reaction between an acid and base (e.g. alkali)

2. Limestone, marble and chalk all contain calcium carbonate. This compound reacts with the acid in rain water and chemical weathering happens. Do some research to find out other rocks and minerals that can react with the acid in rain water.

Did You Know?

You cannot dig up pure iron from the Earth's crust as it is always oxidised to the compound iron oxide. However meteorites that have just landed on Earth can be pure iron because they have not had a chance to oxidise!

KEY WORDS

carbonic acid
neutralisation
chemical weathering
oxidation

Acid Rain

- What is acid rain?
- What causes acid rain?
- What are the effects of acid rain?
- How can we stop acid rain?

Did You Know?

Many petrol stations now only sell low-sulfur fuels. These are fossil fuels which have had the sulfur removed before the fuel is put into the car engine. The sulfur is then sold to the chemical industry where it is used in the Contact Process to make sulfuric acid. Sulfuric acid is needed to make many products including paints and medicines.

What is acid rain?

As we saw on the previous pages, all rain water is weakly acidic. Air **pollution** can cause rain water to have a pH of less than 5.5. This more acidic rain is called **acid rain** and contains strong acids.

a What are the similarities and differences between rain water and acid rain?

Many fossil fuels have sulfur impurities in them. When we burn these fuels in power stations or car engines, the sulfur also combusts to form **sulfur dioxide**. Sulfur dioxide reacts with rain water and oxygen in the air to make sulfuric acid. This lowers the pH of rain water.

Air is a mixture of gases; about 80% is nitrogen and most of the rest is oxygen. Air is pulled into the engine of cars to give oxygen for the combustion of the fuel. In the hot car engine the nitrogen combusts to make **nitrogen oxides**. The nitrogen oxides react with rain water and oxygen to make nitric acid. We find this strong acid in some acid rain.

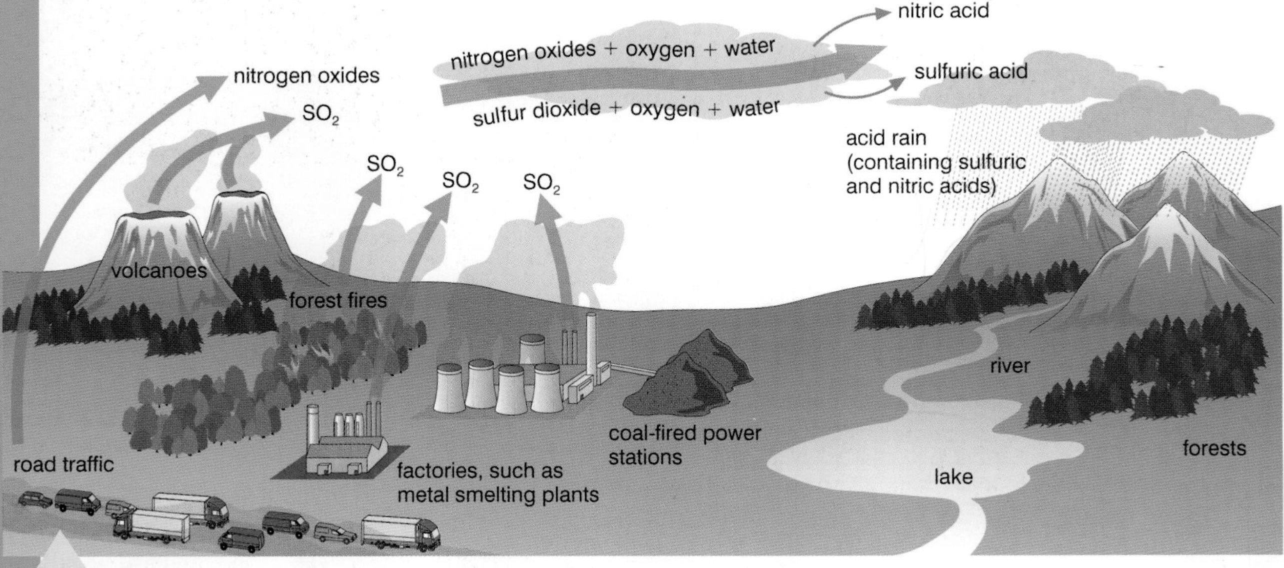

Causes of acid rain

b Which two strong acids can cause acid rain?

Acid rain can have disastrous effects on forests

Pollution problems

Acid rain has a number of effects:

- lowers the pH of lakes and this reduces the amount of life that can live in the water
- damages leaves of plants and reduces their ability to photosynthesise; can kill them
- lowers the pH of farming soil, stopping some crops from growing
- increases chemical weathering of rocks, metals and buildings.

Acid rain can be stopped. The sulfur in fossil fuels can be removed before we burn the fuel. Or the waste gases from burning fossil fuels can be passed through a gas scrubber. This contains a base (pH greater than 7) which neutralises the acidic nitrogen oxides and sulfur dioxide gases. Catalytic converters in car engines change the nitrogen oxide back to nitrogen before the gas is released into the air.

Once the acid rain has fallen the damage can be limited. You can do this by adding weak bases, like powdered limestone into the environment to increase the pH.

Great Debates

Catalytic converters have been fitted to all new UK cars since the 1990s. They are expensive, need to be changed regularly and only work when they are warm. However catalytic converters can help reduce acid rain caused by cars and stop poisonous carbon monoxide and unburnt fuel being released.

Do you think catalytic converters are a good idea? Should they be fitted to all cars, or should the car owner choose whether they have them fitted?

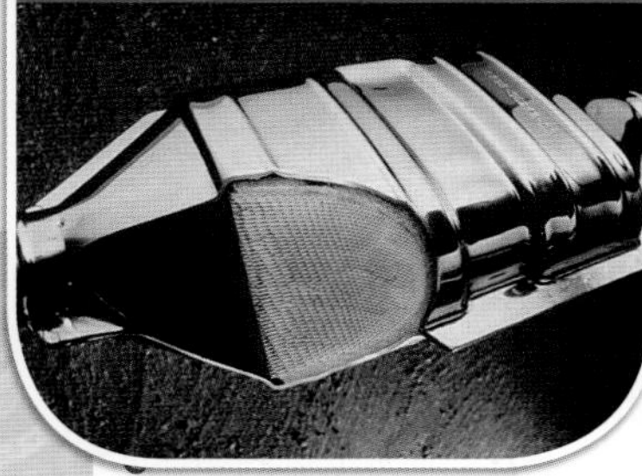

Summary Questions

1. Copy and complete the sentences below:

   Acid rain contains …

   Acid rain can be neutralised …

   Sulfur is an impurity …

   Nitrogen in car engines …

2. Write two word equations to show how sulfur from fossil fuels can make sulfuric acid. Hint: The first step is the combustion of sulfur, the second step is making the sulfuric acid.

3. Write two word equations to show how nitrogen in the air can make nitric acid. Hint: The first step is the combustion of nitrogen; the second step is making the nitric acid.

4. Make a poster to explain what can cause acid rain and how it can be reduced.

KEY WORDS

pollution
acid rain
sulfur dioxide
nitrogen oxides

Physical Weathering

- What is physical weathering?
- How does physical weathering change rocks?

Let's get physical!

Forces caused by nature can break down rocks into smaller pieces. The chemicals in the rock are unchanged. This is **physical weathering**.

a Explain why physical weathering is not a chemical reaction.

Rocks are mixtures of minerals. As the environment changes around rocks, different minerals in the rocks change in different ways. These changes cause the rock to break into smaller pieces, or to be weathered.

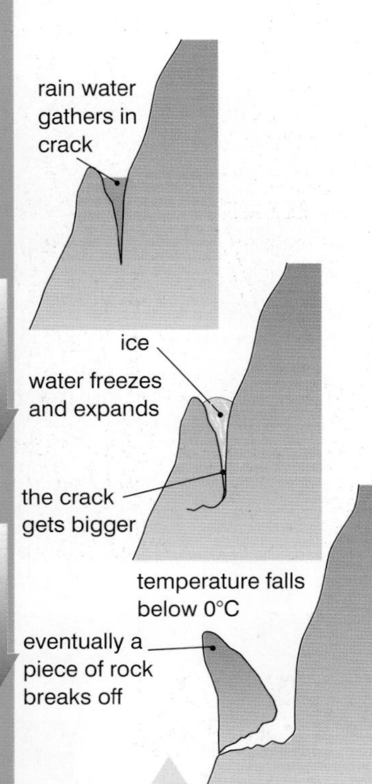
rain water gathers in crack
ice
water freezes and expands
the crack gets bigger
temperature falls below 0°C
eventually a piece of rock breaks off

The effects of freeze–thaw weathering

Ice breaker

Freeze–thaw weathering is caused by rain water trapped in small cracks in rocks. As the temperature drops below the freezing point of water, ice is made. This can break the rock into smaller pieces. What happens when you freeze a can of pop?

b What special property of water allows it to break up rocks when it freezes?

Link up to...

GEOGRAPHY

Freeze–thaw weathering is known as 'frost shattering', and onion-skin weathering is called 'exfoliation'.

activity

Investigating freeze–thaw weathering

Water can get trapped in rocks and as temperatures change, the water can freeze and this can cause weathering. You are going to investigate why water freezing can cause rocks to be weathered physically.

- Measure 100 cm^3 of water into a measuring cylinder.
- Freeze the water in the measuring cylinder.
- Measure the volume of the ice.
- Use this experiment to explain how freeze–thaw weathering can break down rocks.

100 90 80 70 60 50 40 30 20 10 cm^3

Onion-skin weathering

Onion-skin weathering is caused by changes in temperature within a day. At night the temperature of the environment drops and this causes the rock to contract. In the day, temperatures rise and the rock expands. As the rocks are made of different minerals, each one expands and contracts by a different amount causing stress in the rock. The stress forces weaken the surface of the rock and it peels off as layers.

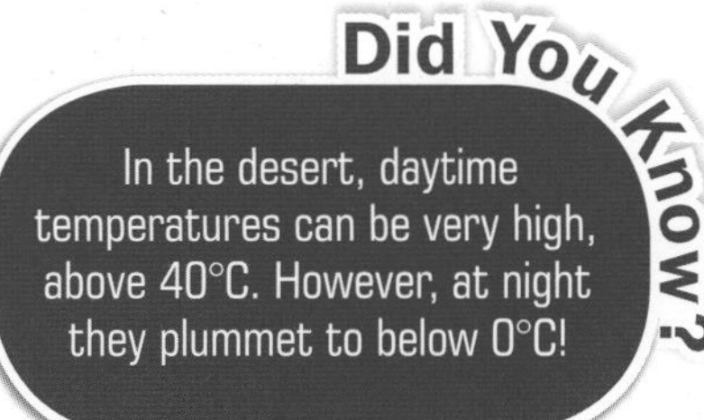
Did You Know?
In the desert, daytime temperatures can be very high, above 40°C. However, at night they plummet to below 0°C!

activity

Demonstrating onion-skin weathering

Changes in temperature can cause rocks to peel off in layers. You are going to investigate how temperature changes over a short space of time can cause rocks to be weathered physically.

- Set up a Bunsen burner.
- Half-fill a beaker with cold water.
- Heat the end of a glass rod in a blue Bunsen flame until the flame turns orange.
- Put the hot end of the glass rod into the water. Observe.
- Re-heat the glass rod and again put it into the water. Observe.
- Use this experiment to explain how onion-skin weathering can break down rocks.

hot glass rod

cold water

Safety: Wear eye protection and be careful not to burn yourself.

Summary Questions

1. Match the start and endings of the following sentences:

a) Physical weathering is	can freeze and expand, causing the crack to get bigger.
b) Large changes in temperature	caused by physical changes in rocks.
c) Trapped rain water in rocks	expand and when they get cold they contract.
d) When rocks get hot, they	can cause rocks to peel.

2. Copy and complete the table to compare biological, chemical and physical weathering:

Information	Biological	Chemical	Physical
Definition			
Type of change			
Human impact			

3. Using the particle model from Fusion 1 Chemistry and Fusion 2 Physics, explain why rocks can expand when they get hot and contract when they get colder. How does this lead to onion-skin weathering?

KEY WORDS

physical weathering
freeze–thaw
onion-skin

Erosion

- What is erosion?
- What happens to weathered rock?

Erosion is the wearing away of rock as it is transported from one place to another. So, erosion is both weathering and transport.

 What do you think could cause erosion?

Moving on

How could this boulder have moved here?
Glaciers covered the U.K. in the Ice Age and they moved large rocks. When they melted the boulders were left.

When rocks have been weathered they can be moved from place to place. This is called **transportation**. Gravity, wind, water or ice can transport the weathered rock to new places.

When rock is transported by water or wind, the rock fragments hit each other and become smaller and more rounded. If transport happens by ice in a glacier, the rock pieces will have more jagged edges. This is because the rock pieces are like cherries in a cake and do not hit each other. However, they are very effective at wearing away the rocks at the bottom and sides of the glacier.

b Look carefully at the pictures of transported rock fragments. Which fragment was transported by ice? How do you know?

c Predict where you would find jagged sediments in a river. Don't forget to explain your prediction using science.

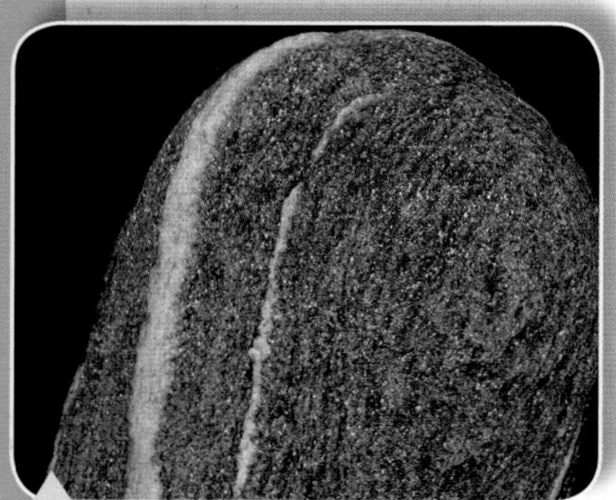

Transported rock fragments A

Transported rock fragments B

If a rock is to be transported, the force of wind, water or ice must be greater than the force of gravity. Small rock pieces are normally transported by the wind.

The rock pieces carried in a river are called 'sediment'. Large, heavy sediment like stones need more **energy** (faster flowing water) to transport them. This large sediment may be rolled along the river bed. Small light particles such as in clay are carried in suspension (float in the water).

The fastest flowing water happens when there has been heavy rain and near the start and centre of the river. When the speed of the water slows down, the sediment stops being transported and the rock settles out of the water or is **deposited**.

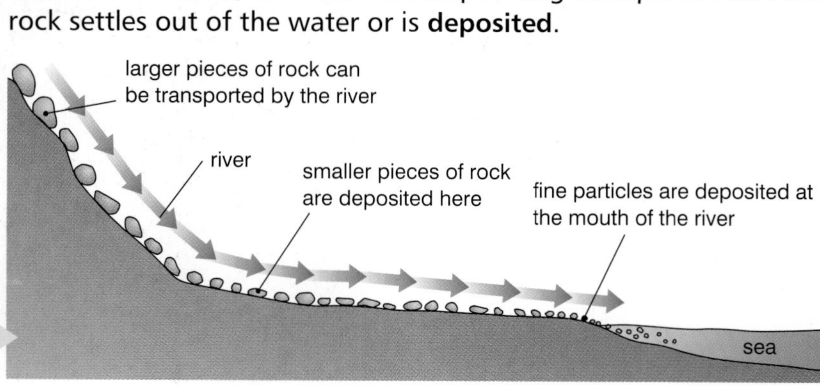

Sediments tend to be sorted by size as the river deposits them along its course to the sea

activity

Modelling a river

Rivers change as you travel from their source (start) to their mouth (where they join the sea). The path of a river changes over time and this can have an effect on where people can live and farm. Environmental scientists study rivers on field trips to help predict how the river bed will change over time. The scientists use their data to suggest ways to manage the river to balance the environmental impact and human impact on the river.

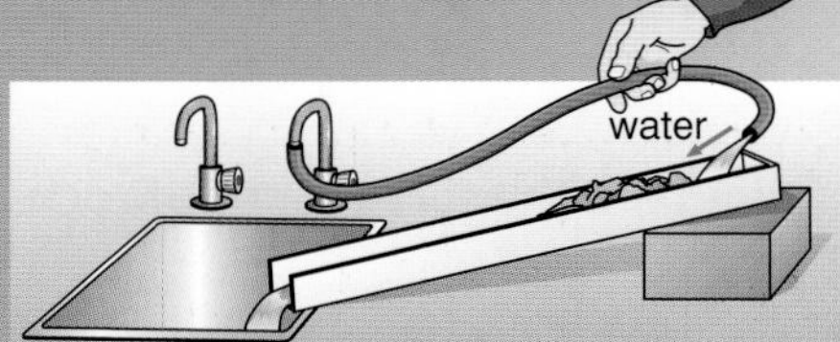

- Look carefully at the model.
- What is representing the river bed?
- What is representing the sediment?
- Where is the energy of the water at its highest and lowest? Why?
- In which part of the model will the largest sediment be transported?
- What are the disadvantages of using this model for transportation?

Did You Know?

Sediments can be carried for thousands of miles and can be deposited at any point where the energy of the river is not great enough to carry it any more. The Yellow River in China deposits about 3 cm depth of sediment every day! Also the Mississippi River deposits about 2 million tonnes of sediment per day, which forms its delta. Sometimes big hydroelectric dams are built upstream to provide electricity. What effects do you think the dam might have on the surrounding land?

Help Yourself

Make sure that you understand the difference between the words 'erosion' and 'weathering'. Try to write a sentence using each of these words correctly.

These pieces of limestone at the bottom of the cliff have been eroded

Summary Questions

1. Match up the key words with their correct definitions:

a) transportation	needed to transport rock, measured in joules
b) deposition	moving pieces of rock
c) energy	wearing away of rock as it is transported
d) erosion	when transported rock is no longer being moved

2. Bob is from a Welsh university and went on a field trip to Snowdonia to survey the size of the sediment in a river.

   At 1 km from the source, he found that the average diameter of the rock was 9.7 mm, at 50 km from the source the average diameter was 1.2 mm and 100 km from the source the average diameter was 0.02 mm.

   a) What is Bob's dependent variable?
   b) What is the unit of Bob's independent variable?
   c) How could Bob display his results in a more scientific way?

KEY WORDS

erosion
transportation
energy
deposited

Rock Cycle

- How can we make new rocks from old?
- What is the rock cycle?

Changes to a weathered coastline

When we look at the countryside around us, it is easy to think that the rocks do not change. However, weathering is always breaking rocks down into smaller pieces. These pieces are transported and deposited. Over many years they become cemented together to make new sedimentary rocks.

Other new rocks can be made when magma comes up from the Earth's crust to make igneous rocks. Heat and pressure can change rocks into metamorphic rocks.

a What are the three types of rock?

The chemicals in rocks are called 'minerals'. Minerals are constantly changing and are being recycled. Sometimes the rocks are just changed in size; other times the minerals in the rocks undergo chemical reactions to form new substances. These changes are too slow to notice daily, but over your lifetime you may see changes in your local landscape.

All the chemicals on earth are recycled over and over again. The **rock cycle** describes how the substances in the rocks are recycled.

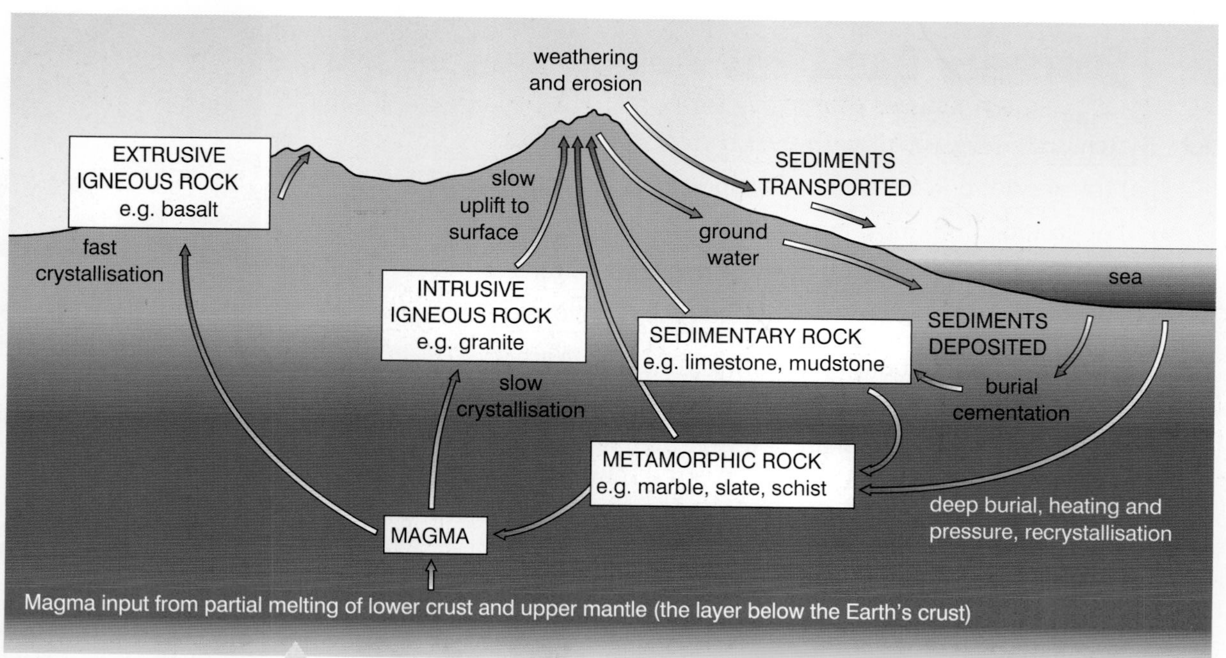

The recycling of rocks can be shown as a flow chart

All rocks are **recycled**.

- Igneous rocks are made when magma cools down and solidifies.
- Igneous rocks are then weathered and eroded to make sediment.
- The sediment is deposited, compacted and cemented to make sedimentary rocks.
- Sedimentary rocks can be changed by heat and/or pressure and this makes metamorphic rocks.
- Movements of the Earth's crust, such as when an earthquake happens, can force rocks deep into the Earth. The rocks are then melted and make magma again and so the rocks are recycled.

Did You Know?

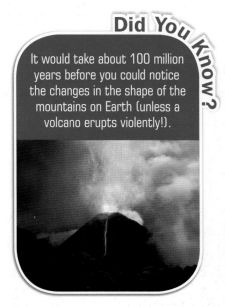

It would take about 100 million years before you could notice the changes in the shape of the mountains on Earth (unless a volcano erupts violently!).

The rock cycle!

b What are the processes that make sedimentary rock?

Summary Questions

1. Copy and complete the sentences below:

   Igneous rocks can become …

   Metamorphic rocks are made …

   The Earth is a …

   Rocks are constantly …

2. What are the differences and similarities between intrusive igneous rock and extrusive igneous rock?

3. Do you think rocks are renewable or non-renewable? Discuss this question with other pupils in your class.

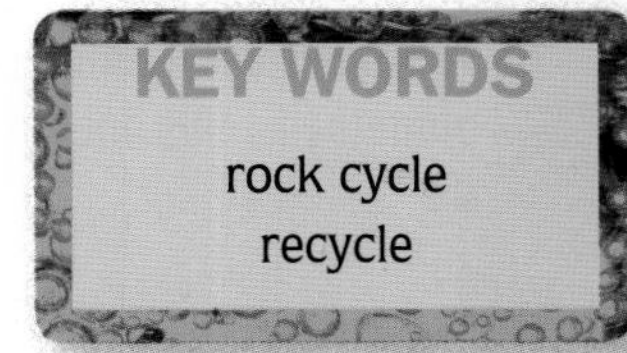

Rocks in the Universe

- Are igneous, sedimentary and metamorphic rocks found elsewhere in the universe?
- Do other planets have rock cycles?

Many scientists believe that other **planets** in our solar system have a rock cycle. For rocks to be recycled, the planet must have erosion. Wind or water is needed for erosion to take place. Rocks that are eroded can form sedimentary rocks. The planets might also have evidence of volcanoes which lead to the formation of igneous rock and metamorphic rocks.

Meteorites regularly land on Earth. As the rocks from space enter the atmosphere they get very hot because of the friction. The heat can make some of the meteorites burn away, looking like streaks of light in the sky. These burning meteorites are called shooting stars. Sometimes the meteorites survive and land on the surface of the Earth, making a crater.

Did You Know?

Is there life on Mars? In 1996 a meterorite landed on Earth, believed to be made from Martian rock. Many scientists believe that there was evidence of microorganisms, a form of life, in the rock sample. Without water, life as we know it could not exist. Then on July 4th 1997, Mars Pathfinder landed in an ancient flood plain, suggesting the presence of water. So, there might have been life on Mars – further space probes have been launched to try to answer this question.

Crater made by a meteorite

a What processes do scientists look for to prove a planet has a rock cycle?

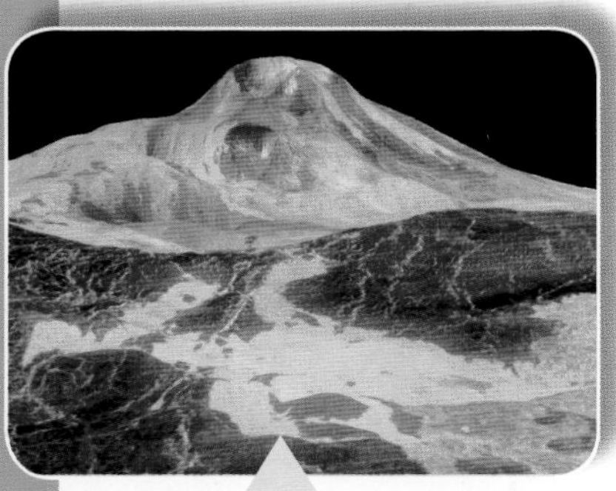

Volcanoes occur on other planets like this one on Venus

Mercury, Venus and Mars are the closest planets to the Sun and known as the **rocky planets**. These planets show evidence to support the idea that they have a rock cycle. However, planets further from the Sun like Jupiter, Saturn, Uranus and Neptune do not show evidence of a rock cycle. The outer planets of our solar system are called **gas giants**.

b What are gas giants?

Evidence for planetary rock cycles

- The Mariner 10 spacecraft has taken thousands of images of Mercury and has shown evidence of lava flows.
- Mars has large volcanoes, for example Olympus Mons.
- Radar mapping has been done for Venus and there have been mountain ranges identified. These seem to have formed as a result of movement of large sections of crust.

Other objects in space also have evidence of a rock cycle. Earth's moon has rock recycling because of the action of solar weathering and gravity.

Rock cycles on other planets and moons happen at a much slower rate than on Earth. This is because the movements in their crust are even slower than those on Earth and there is not as much erosion.

1. Match up the key words with their definitions:

a) planet	planets in the outer part of our solar system
b) rocky planets	a massive object that orbits a star
c) gas giants	the planets closest to the Sun

2. Choose an object in the solar system and research it. Find out what data has been collected to show if it has or has not got a rock cycle. Present your findings in a PowerPoint® slide show.

3. Why do you think rock cycles are slower on other planets? Discuss this question with other pupils in your class.

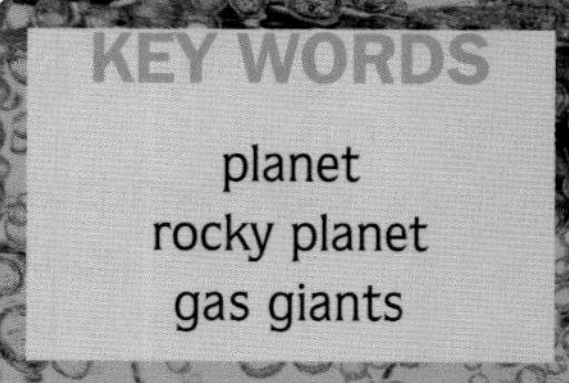

know your stuff

▼ Question 1 (level 4)

Rocks can be put into three groups: sedimentary, metamorphic and igneous.

a Match up the type of rock with how it is made.

sedimentary	made from solidified magma.
igneous	rocks that have been changed by heat and/or pressure.
metamorphic	made from eroded rocks that are compacted and cemented together.

[3]

b Which of the following is *not* a sedimentary rock?

chalk limestone marble [1]

c Rain is slightly acidic and causes weathering of chalk, limestone and marble. What type of weathering is this? Choose from the list below.

biological chemical physical [1]

▼ Question 2 (level 5)

Below is a picture of Jurassic sandstone found in the USA. Look carefully at the picture and you can see the layers or beds in the rock.

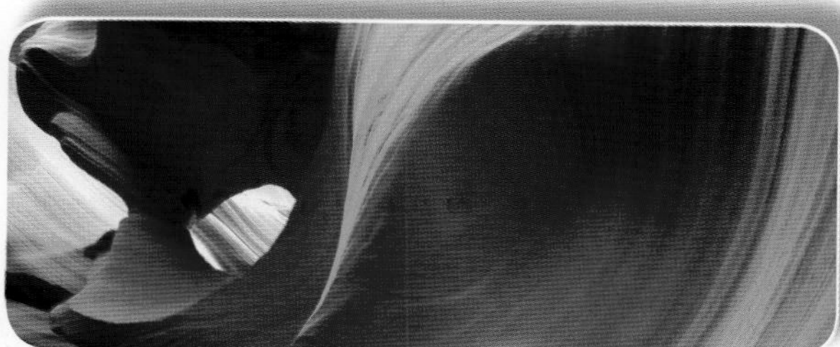

a What type of rock is sandstone? [1]

b Sandstone absorbs water and is known as 'porous'. What is the grain structure in sandstone? [1]

c Explain how sandstone is formed. [4]

▼ Question 3 (level 6)

Marble is a rock that contains mainly calcium carbonate:

a What are the names of the elements that are in calcium carbonate? [1]

b What type of rock is marble? [1]

c Explain how marble is made. [2]

d Acid rain can contain nitric acid. Complete the word equation for the reaction between calcium carbonate and nitric acid:

calcium carbonate + nitric acid ⟶ calcium nitrate + … + … [2]

e What type of weathering is this? [1]

▼ Question 4 (level 7)

The rock cycle is a model which is used to explain how rocks on earth are recycled.

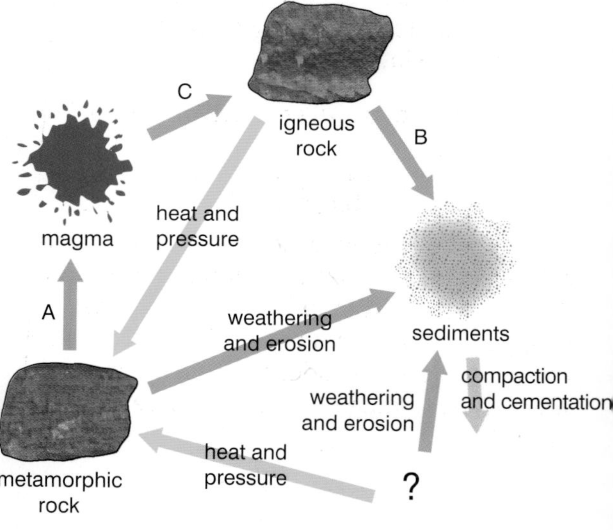

a Which type of rock is missing from the diagram? [1]

b Name the processes A, B and C. [3]

c What is erosion? [2]

How Science Works

▼ Question 1 (level 4)

Doug and Mohammed wanted to investigate if temperature affected the speed of chemical weathering. They measured the time it took for a piece of marble to produce 20 cm^3 of gas when it was in acid at 0°C, 10°C, 20°C, 30°C and 40°C.

a What are the control variables in Doug and Mohammed's experiment? These are the variables they kept constant to make sure this was a fair test. [2]

b Explain how Doug and Mohammed could cool acid down to 0°C. [1]

c Which of the following pieces of equipment would be the most sensitive way to measure the volume of acid? Choose one from the list below.

beaker measuring jug
measuring cylinder balance [1]

d What piece of equipment would Doug and Mohammed use to measure the independent variable (the variable they chose to investigate)?

thermometer ruler stopwatch balance [1]

▼ Question 2 (level 5)

Louise decided to investigate how temperature affects the crystal size in an igneous rock. She used liquid salol to model magma. Louise dropped a small amount of salol on warmed or cooled microscope slides and let it solidify. She then measured the diameter of three crystals on each slide.

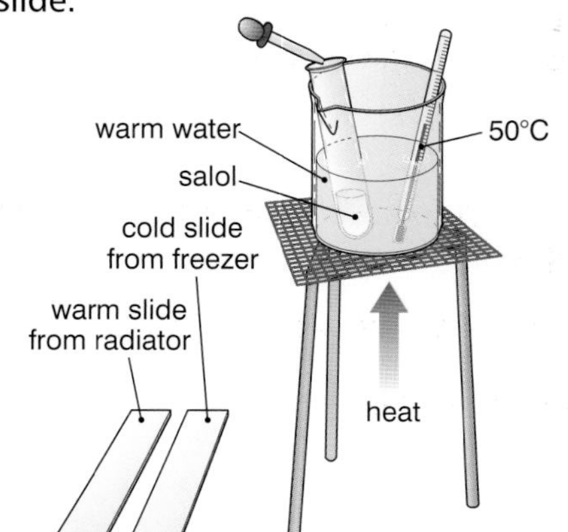

a Why did Louise have to use a model to investigate the crystal size in rocks? [1]

b What is the independent variable in Louise's experiment? (The independent variable is the one she chose to investigate.) [1]

c What is the dependent variable in Louise's experiment? (The dependent variable is the one she measured to see the effect of changing the independent variable.) [1]

d How could Louise make her measurements more precise? [1]

▼ Question 3 (level 6)

Dan and Charlie went on a field trip and collected information about the average sediment size in a river. They found out that at 0 km from the source of the river, the rocks had an average diameter of 10.1 mm, at 5 km it was 9.3 mm, at 15 km it was 5.1 mm, at 10 km it was 8 mm and at 20 km it was 4.4 mm.

a Draw a table of Dan and Charlie's results. [4]

b Which is the least precise measurement of diameter? [1]

c Explain why a line graph could be drawn from this data. [2]

P1.1 Light and Space

Light from space

The squid in the photograph live in the depths of the ocean where it is very, very dark. The Sun's rays cannot reach that far down but the squid can still see each other. That's because they glow in the dark. They have special organs where chemical reactions produce light.

We are different from the squid. Where we live, we have the Sun to light up the world around us. Every day, the Sun rises in the East and sets in the West. Even when it is hidden by clouds, it provides us with the light we need to see.

At night, we make use of artificial sources of light, particularly electric lighting; and at night we can see the stars. If you know where to look, you can even see some of the other planets of the solar system.

In this topic, you will build on what you already know about light and space. You will understand more about the scientific picture of light. You will also see how it has helped us to learn more about everything that is out in space.

Glow-in-the-dark squid

 Look at the photo of the squid. Why is it so dark where they live? Why do you think they need to glow like this?

activity

Seeing the light

Watch as your teacher demonstrates how the light from a laser travels. Laser light is ordinary light, but it comes out in a narrow, bright beam. This makes it easy to see how it travels.

- Find out why we can only see the laser beam when there is dust in the air or when the water is cloudy.
- What happens when the beam reaches a surface which reflects it?

Safety: Your teacher will explain the precautions to ensure that no-one gets the laser beam in their eye. A bright laser beam can cause permanent damage to the eye, including blindness.

Light ideas

Look around the room. Close your eyes and you won't be able to see. Here are two explanations for this:

- Your eye sends out 'feelers' of light into the room. These feelers sense the objects in the room, so that you know that they are there. When you close your eyes, light from your eyes can no longer feel the room.
- The objects in the room give off light. When this light enters your eyes, you can see the objects. When you close your eyes, no light enters them and you cannot see.

Discuss with a partner which is the better idea.

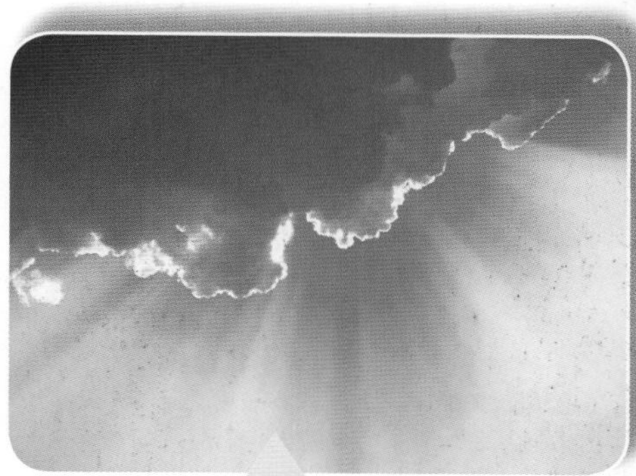

The Sun may be hidden by clouds, but you still know that it's there

For a long time people tho ught the first idea might be right. However, the second idea is the modern scientific way of thinking about how light allows us to see.

When there are two different ideas to explain something, scientists try to think of an experiment to test them to find out which is right.

activity

Sun safety

You should never look directly at the Sun, or through binoculars or a telescope. Later in this unit, you will learn a safe way to see the surface of the Sun.

If you stare at the Sun, its light can blind you. However, we can often see the Sun in the sky, and yet we are not blinded. Use the Internet to find out how our bodies automatically react to bright light to protect us from danger.

Look at the photo of sunlight coming from behind a cloud. Use what you already know about light to answer these questions:

- Why can't we see the Sun in the photo? Why are the edges of the cloud brightly lit while the middle is dark?
- Make a sketch of the photo and indicate where you think the Sun is. Explain how you know.
- Usually, when we see the Sun in the sky, we cannot see the beams of light coming from it. Why can we see them in the photo? Why are they usually invisible?

Straight-line Light

- How do we know that light travels in straight lines?
- How fast does light travel?

Seeing straight

The surveyor in the photograph is working on a building site. The tall instrument next to him is a 'theodolite'. He uses this to check that the sides of the building are built in straight lines. He makes sure that the walls meet perfectly at the corners.

In the past, builders used lengths of string, stretched as tightly as possible between two posts, to give a straight line. A stretched string always sags slightly in the middle. The theodolite is better – it uses a laser beam to give a perfect straight line. A beam of light doesn't sag!

Using a theodolite on a building site

a The theodolite relies on light travelling in straight lines. Look back to the photos on the previous two pages. What evidence can you see to support the idea that light travels in straight lines?

activity

Light line-up

Here are two ways to test the idea that light travels in straight lines. You will need several pieces of card *and* wooden blocks to support them vertically.

Method 1

- Use a pin to make a hole in the centre of each of three cards.
- Pass a length of thread through the three holes. Stand the cards vertically so that they are separated by a few centimetres.
- Pull the thread taut so that the holes in the cards are in a straight line.
- Now remove the thread (but don't disturb the positions of the cards).
- Shine a light through one hole; does light pass through all three holes?
- Move the middle card a small distance to the side; what happens?

Method 2

Your teacher will show you how to use a ray box to produce a narrow, straight ray of light on a piece of white paper. With a pencil, mark three dots on the ray. Switch off the light and use a ruler to check whether the three dots lie in a straight line.

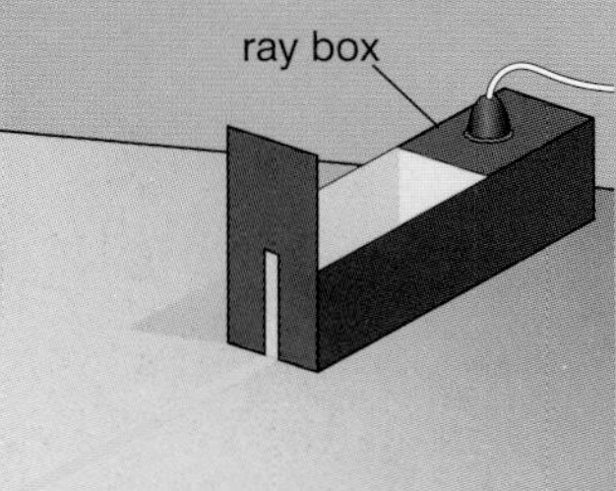

Zapper test

When you change the channel on the TV, you must point the remote control towards the TV set. The remote control uses infra-red radiation which we cannot see. (Some remote control toy cars also work with infra-red.)

Your task is to devise a way of finding out whether infra-red radiation travels in straight lines, just like light.

Ray comes from here

Faster than a speeding bullet

Light travels fast. It travels through space at 300 000 kilometres per second (km/s). At this speed, it could travel 7 times round the Earth in a second.

The Sun is 150 million kilometres away. It takes light from the Sun 500 s to get here – that's about 8 minutes. If the Sun suddenly stopped shining, it would be 8 minutes before we knew anything about it.

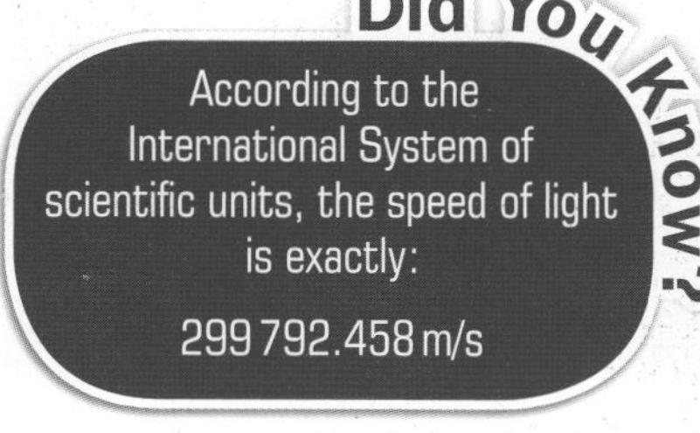

Did You Know?
According to the International System of scientific units, the speed of light is exactly:
299 792.458 m/s

The **speed of light** is a very special speed. It seems that nothing can go faster than light. Some scientists think that if we could find a way of moving faster than the speed of light, we would end up travelling backwards in time. That could change history lessons in a big way!

Last call for passengers to see the Magna Carta signed.

Summary Questions

1. We can use a ruler to show the path of a ray of light. Why is a ruler good for this?

2. The speed of light is 300 000 km/s. The Moon is 400 000 km from the Earth. How long will it take light from the Moon to reach the Earth? Choose one from the list below:

   less than a second **between 1 and 10 seconds**
   more than 10 seconds

Materials and Light

- What happens when light hits different materials?
- How can we measure the brightness of light?

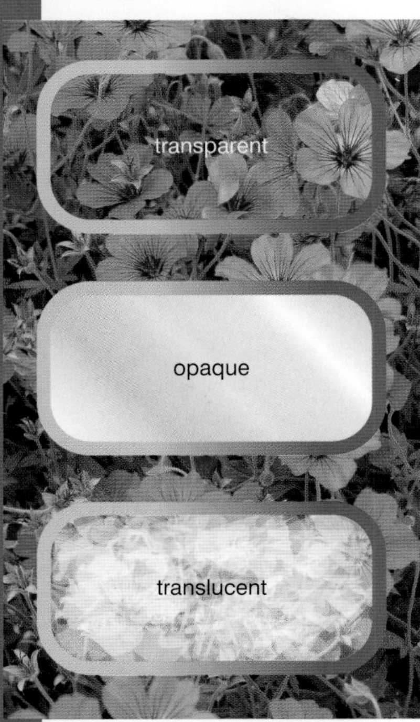

Describing materials

Look out of the window. You can see straight through the pane of glass in the window. We say that glass is a transparent material, because light passes straight through it.

If the window is broken, it may be replaced temporarily with a wooden board. You can't see through wood – it is an opaque material.

You don't want anyone peering in the bathroom window, so it may be covered by a blind (opaque) or the glass in the window may be translucent. A translucent material lets light through but you cannot see a clear image through it.

a Window glass is usually transparent. Name two other transparent materials. Can a transparent material be coloured?

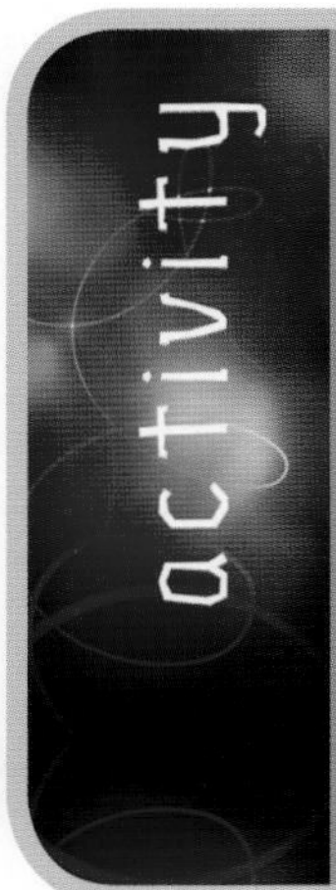

Examining materials

Polymers (plastics) can be transparent, translucent or opaque. Examine some different polymers and decide how to classify them.

activity

Measuring light

A light meter can tell you how bright the light is. Lay the meter down on the bench, facing upwards. Fix a lamp or torch above it, shining downwards on to the sensor of the meter. Note the reading.

Now put a sheet of polymer between the lamp and the meter. Does the reading change?

How can you use the meter to compare different materials? How can you make this a fair test?

A light meter has a sensor to detect the light and a scale to show the brightness of the light

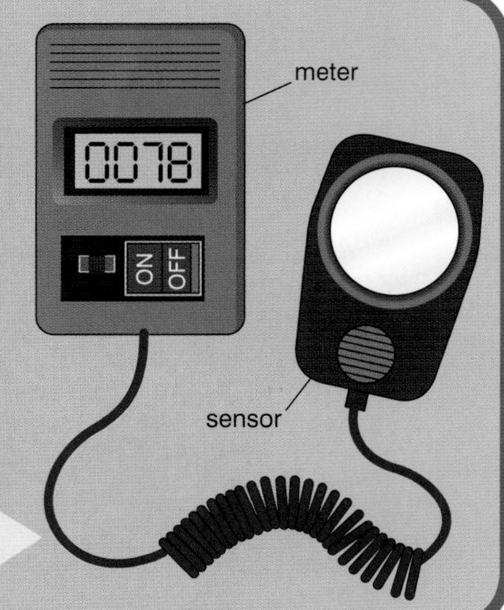

Absorb, transmit, reflect

Why are some materials transparent while others are opaque? What happens when light passes through a translucent material? The diagrams show what happens to rays of light when they strike different materials.

- When rays strike a transparent material, they pass straight through unaffected. The material **transmits** light.
- Some opaque materials **absorb** light. No light passes through to the other side.
- Some opaque materials **reflect** the light that falls on them.
- As light passes through a translucent material, it is **scattered** in all directions.

b Which diagrams show transparent and translucent materials? How can you tell from the diagrams that light passes through these materials?

Summary Questions

1. Copy the table and complete it by writing a single word in each space in the first column.

	describes a material that doesn't allow light to pass through
	describes a material that allows rays of light to pass straight through
	describes a material that can be seen through but not clearly

2. a) Why might a front door be fitted with a panel of translucent glass?
   b) Suggest some other uses for translucent materials.

3. Imagine shining a torch beam on a sheet of matt black paper. No light is transmitted through the paper. None is reflected by the paper. So what happens to the light?

4. A sheet of ice can look perfectly transparent. Describe an experiment you would perform to find out if 100% of light is transmitted by ice.

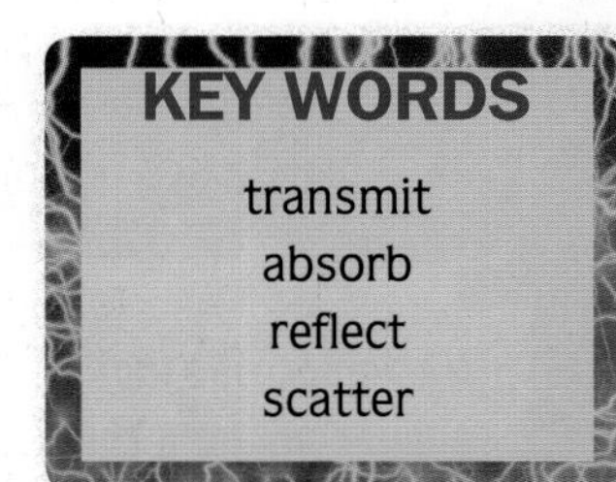

Mirror, Mirror on the Wall...

P1.4

- What is happening when we see a reflection in a mirror?
- How can we predict where a reflected ray of light will go?

A kaleidoscope that's big enough for a child

On reflection

Look around you – most of the things you see are opaque. You can see them because they reflect light to your eyes. A mirror is an almost perfect reflector. We can use mirrors to find out more about how light is reflected.

A plane (flat) mirror has an almost perfectly smooth, flat surface. When you look into a mirror, you see an almost perfect reflection of yourself. We say that you are seeing an **image** of yourself.

The man in the photo is holding his tie with his left hand. However, his image seems to be using his right hand. This shows that the image in a plane mirror is 'left-right' reversed. Here is what we can say about the image formed by a plane mirror:

- The image is the same size as the original object.
- The image is as far behind the mirror as the object is in front.
- The image is inverted ('left-right' reversed).

The law of reflection

The diagram shows the path followed by a ray of light when it reflects off a plane mirror.

The two rays are called the **incident ray** and the **reflected ray**.

The **normal** is the line drawn at right angles (90°) to the mirror at the point where the incident ray strikes it.

The angle of incidence is the angle between the incident ray and the normal. The angle of reflection is the angle between the reflected ray and the normal.

The law of reflection says that these two angles are equal:

angle of incidence = angle of reflection

This law is important because it allows us to predict the path of a ray of light when it strikes a mirror or any other reflecting surface.

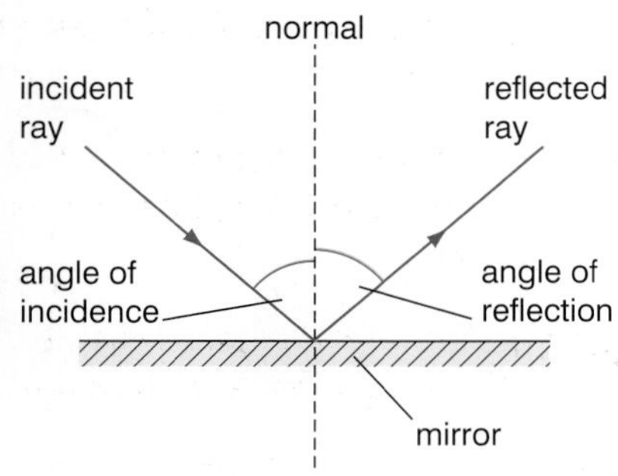

When a ray of light reflects off a mirror, the angle of incidence is equal to the angle of reflection

a Imagine a ray of light striking a mirror with an angle of incidence of 60°. What will the angle of reflection be?

activity

Investigating reflection

- Use a wooden block as a stand for a plane mirror on a sheet of white paper.
- Shine a ray of light from a ray box at the mirror. Shine it at an angle and look for the reflected ray.
- On the paper, mark the positions of the mirror and of the incident and reflected rays.
- Remove the mirror and ray box; draw the normal to the mirror.
- Measure the angles of incidence and reflection; are they equal?
- Try again for a different angle of incidence.

Forming an image

If you look into a mirror, why does your image seem to be behind it? The diagram helps to explain this.

Suppose you look at the reflection of the tip of your nose. Rays of light from your nose reflect off the mirror, as shown. The dashed lines show that they *appear* to be coming from a point behind the mirror. *So* your mind imagines that there is a nose there.

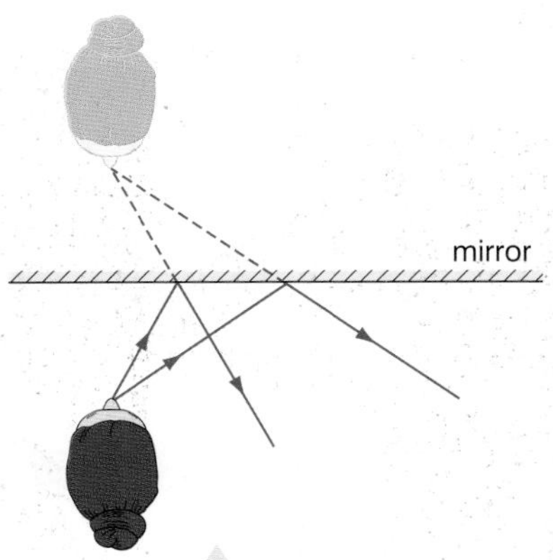

Why you think your twin is behind the mirror

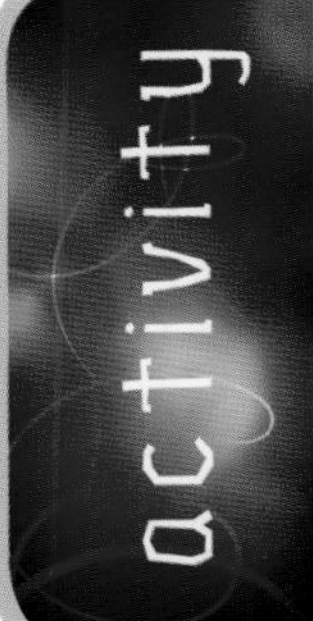

activity

Multi-mirror miracles

You can use a periscope to see around a corner or over an obstacle.

- Try to make a periscope using two mirrors.
- You can also make a kaleidoscope using two mirrors.

Next time you...

... get in a car, ask the driver how he or she uses the mirrors.

Summary Questions

1. What is the angle between the normal and the surface of a mirror?
2. Using a ruler and a protractor, draw an accurate diagram showing a ray of light striking a plane mirror with an angle of incidence of 30°. Show the normal and the reflected ray and mark the two angles which are equal (according to the law of reflection).
3. Look at the diagram which shows how the image is formed by a plane mirror. Use the diagram to explain why we say that the image is as far behind the mirror as the object is in front.
4. Look at the photo of the boy in the kaleidoscope on the opposite page. How many mirrors are there? What is the angle between the mirrors?
5. What is a *plane* mirror? Find out about the different shapes of curved mirrors. Which shape produces a magnified (enlarged) image?

KEY WORDS

image
incident
reflected
normal

Rays that Bend

▸▸ What happens to rays of light when they pass through a glass block?

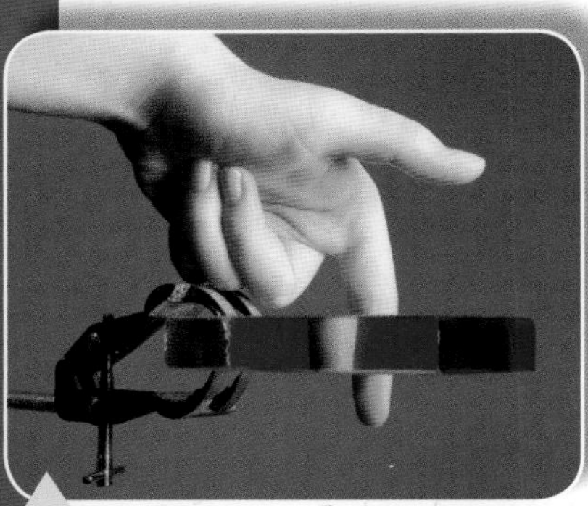

Seeing through glass – an illusion caused by the refraction of light

The 'broken straw' effect

Light illusions

The photograph shows what happens when you look at a finger through a thick block of glass. A section of the finger seems to have been cut out and moved to the side – not very nice!

Bending pencils, disappearing coins

Strange things can happen when light moves between air, water and glass.

- Watch some illusions.
- Can you work out what is going on?

Refraction of light

The strange effects you have seen come about because light is travelling through different materials – air, water and glass. To understand better what is going on, look at what happens when a ray of light goes into a glass block and comes out again.

In the diagram, the **normal** (90°) line is shown at the points where the ray of light enters the glass block and where it leaves it again. You should be able to see that:

- Where the ray enters the glass, it bends *towards* the normal.
- Where the ray leaves the glass, it bends *away* from the normal.

This bending of light when it passes from one material to another is called **refraction**.

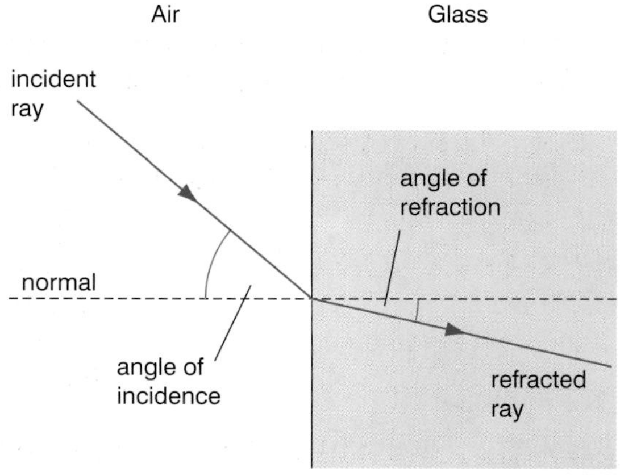

Help Yourself

Remember that light travels in a straight line. It only changes direction at the point where it enters or leaves the glass.

a When a ray of light enters a glass block, which angle is bigger, the angle of incidence or the angle of refraction?

Investigating refraction

- Collect a ray box and a rectangular glass block.
- Place them on a sheet of white paper so that you can mark where the ray goes. Draw around the block to mark its position.
- Shine a ray into the block so that it passes through the glass and out of the opposite side. Mark the path of the ray of light.
- Remove the equipment from the paper and draw in the normal.
- Measure the angles of incidence and refraction. What do you find?
- Now find out what happens when a ray of light enters the glass block at 90° (i.e. along the normal).

Safety: Beware of sharp edges on the glass block.

The shallow pool

If you look down into a pool of water, it could look quite shallow. But beware! It may be deeper than you think. The diagram shows why.

If you see a coin lying on the bottom of the pool, rays of light coming from the coin are entering your eye. The rays change direction as they leave the water. The dashed lines show that the rays *appear* to be coming from a coin higher up in the water, so the pool looks shallower than it really is. Think twice before you jump in.

Summary Questions

1. What do we call the bending of light when it passes from one material to another?
2. When a ray of light travels from air into water, which way does it bend – towards the normal or away from the normal?
3. Copy and complete the diagram to show what happens when a ray of light passes from glass into air.

   glass | air

   a) Draw and label the normal to the surface.
   b) Label the incident and refracted rays.
   c) Label the angles of incidence and refraction.
4. Our eyes make use of refraction. Do some research to find a diagram showing how rays of light are bent as they enter the eye, so that they form an image on the retina.

Did You Know?

Some fishes have specially designed eyes. One half is above the water, the other below. Their brains take account of the effects of refraction.

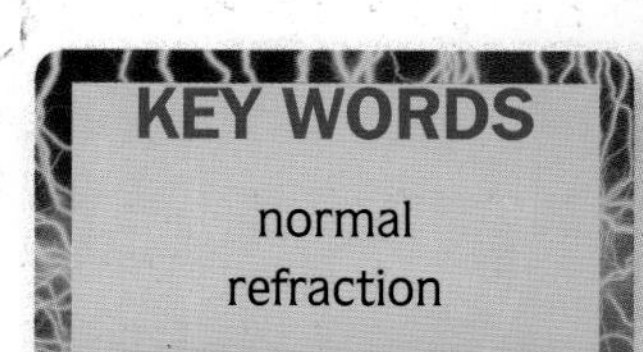

Colours of the Rainbow

- How can we show all the colours which make up white light?
- How do coloured filters affect white light?
- Why does an object change its appearance under coloured light?

Splitting light

If you look at the glass prisms of a chandelier, you will see bright colours. The glass is colourless but we see all the colours of the rainbow. How can we explain this?

For a long time, people thought that the glass gave the colours to the light. However, Isaac Newton realised that the colours were already there in the light, before it struck the glass. The glass simply separates out the different colours.

Here is how he showed this: he shone white light at a glass **prism**. The result was a **spectrum** – all the colours of the rainbow.

This was not enough; it could still be that the glass prism was adding the colours to the light. So he passed the spectrum through another prism, the opposite way round to the first one. The colours recombined to make white light.

This showed that white light is a mixture of all the colours of the spectrum:

red, orange, yellow, green, blue, indigo, violet

The spectrum appears because the prism refracts the light. Violet light is refracted the most, red the least.

A prism splits white light into a spectrum

a Which colour of light is refracted more, green or blue?

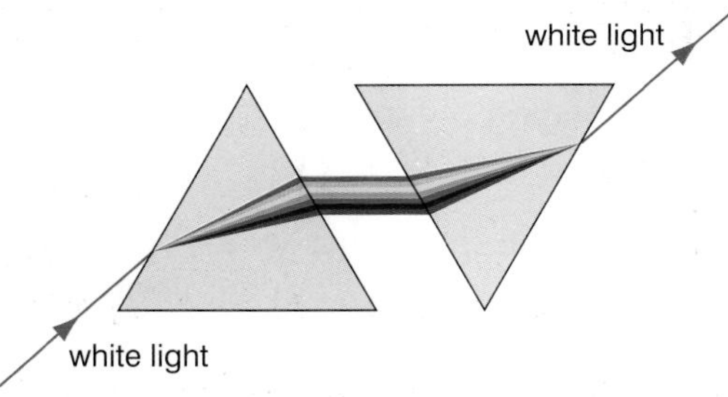

Isaac Newton's idea – recombining the colours of the spectrum to make white light

Did You Know?

You might expect the different coloured bands in the spectrum to have equal widths, but they don't. Yellow is the narrowest. That's an effect of our eyes and brains.

activity

Making spectra

Watch what happens when white light and coloured light are passed through a prism.

How filters work

A coloured **filter** is a piece of transparent coloured material. If you shine white light through a red filter, you get red light.

Does the filter transmit or absorb light? The answer is, a bit of both.

A red filter transmits red light. It absorbs the other colours of the spectrum.

b Which colour of light does a blue filter transmit? Which colours does it absorb?

Using filters

If you look at a white sheet of paper through a red filter, it looks red. White light from the paper reaches the filter; only red light is transmitted, so you see red.

Now, if you look at a blue object through a red filter, what happens? The light coming from the object is blue. It reaches the filter and is absorbed – a red filter will not transmit blue light. So no light reaches your eye from the object and it looks black.

The right-hand half of this photo is seen through a red filter

activity

Hidden messages

- View some coloured objects through coloured filters.
- Record what you see.
- Write a secret message which can only be read through a filter.

Summary Questions

1. Name the colours of the spectrum in order, starting with violet.
2. When white light is shone through a yellow filter, what colour is the light that passes through? What happens to the other colours?
3. A beam of white light strikes two filters: the first one is red and the second is green. What effect will this have on the light?
4. If you look at a green object through a blue filter, what colour will it appear? Explain your answer.
5. If white light is shone through a prism, a spectrum is produced. What happens if red light is shone through the same prism?

Sources of Light

- Which objects in space are sources of light?
- Which do we see because they reflect light?

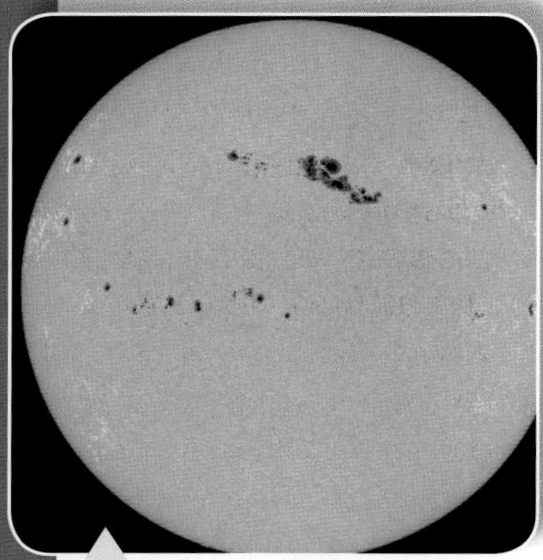

This photo of the Sun's surface was taken by a camera on board a spacecraft

Sun, stars and supernovae

The biggest, brightest thing in the sky is the Sun – at least during the daytime. The Sun is a brilliant source of light because it is a very hot object. Its surface is at a temperature of about 5500°C.

Hot objects give out light. The hotter they are, the brighter they get. The Sun may be 150 million kilometres away, but it is still bright enough to flood the Earth with light.

It is difficult and dangerous to look directly at the Sun. We have an automatic reflex to look away from such a bright source of light. If you could see the Sun's surface, you would notice some dark patches on it. These are sunspots. A sunspot is a cooler region on the surface of the Sun – but they are not that cool, perhaps 4000°C.

The Sun is a **star**. Because the Sun is so bright, we cannot see any other stars during the daytime. Each tiny point of light in the night sky is a star like our Sun, far, far off in space.

activity

Seeing sunspots

There is a safe way to see the surface of the Sun. Use binoculars to project an image of the Sun onto a screen.

With a sheet of paper fixed over the screen, you can draw an outline of the Sun and mark on any sunspots which are visible.

If you repeat these observations over a week or two, you should be able to see the sunspots move across the face of the Sun as it slowly rotates.

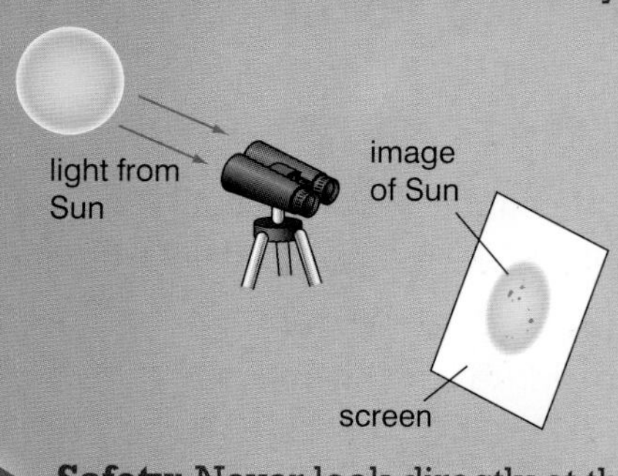

Safety: Never look directly at the Sun, especially when using binoculars or telescopes.

Did You Know?

Uranus is an unusual planet, it rotates very differently to the other planets in the solar system with its poles angled towards or away from the Sun; a 98° degree 'tilt' compared to Earth's 23°. This means that its poles get 42 years of continuous sunlight followed by 42 years of continual darkness as the planet travels around the Sun over its 84 year long orbit.

a Name a hot object in the home which is hot enough to give out light.

On reflection

We can see the Moon although it is not a hot object like the Sun. In fact, the Moon is slightly cooler than the Earth. We see the Moon because it reflects light from the Sun.

We also see the other **planets** by reflected light. None of them is hot enough to be the source of its own light.

The photo shows a spacecraft visiting Jupiter. You can see that only one side of the planet is lit up – the side that is facing towards the Sun.

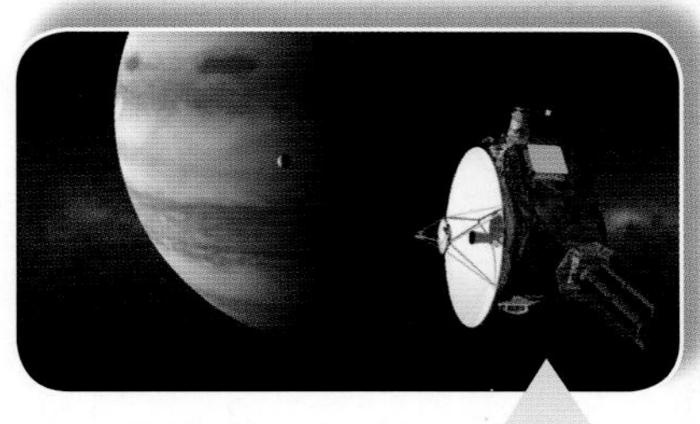

The New Horizons spacecraft in orbit around Jupiter

b Look at the photograph of Jupiter. In which direction is the Sun?

We experience night and day because, at any one time, only half of the Earth's surface is lit up. The other half is in shadow – it's night-time. As the Earth rotates, we move in and out of darkness.

Help Yourself

Do you remember what happens when light falls on an object? Planets are opaque objects. They absorb some of the light that falls on them, and reflect the rest.

Day and night on Planet Football

- Hang a football on a string, or place it on a stand so that it can rotate.
- Shine a light on the ball from one side.
- Use this arrangement to explain why we experience night and day.

Summary Questions

1. Which of the following objects do we see because they are sources of light? Which do we see because they reflect light? Make two lists.

   Sun Moon stars planets

2. A comet is a lump of rock and ice, orbiting around the Sun. Explain how we can see such a cold object.

3. The diagram shows a cross-section of a crater on the Moon.
   a) Copy the diagram and add rays of sunlight falling on the crater. Which part of the crater will be in shadow?
   b) Find some photographs of the Moon's surface showing craters. Can you explain the craters' shadows?

4. The photograph of Jupiter at the top of this page is not genuine – it is made from two separate photographs. Explain why this must be so.

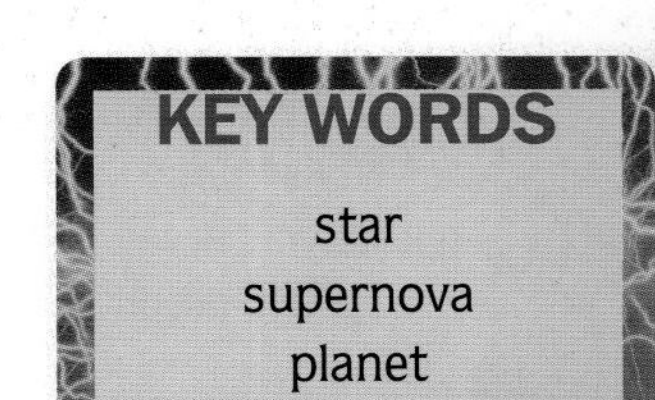

KEY WORDS

star
supernova
planet

Solar System

P1.8

- How do scientists learn about the planets?
- What would it be like to visit another planet?

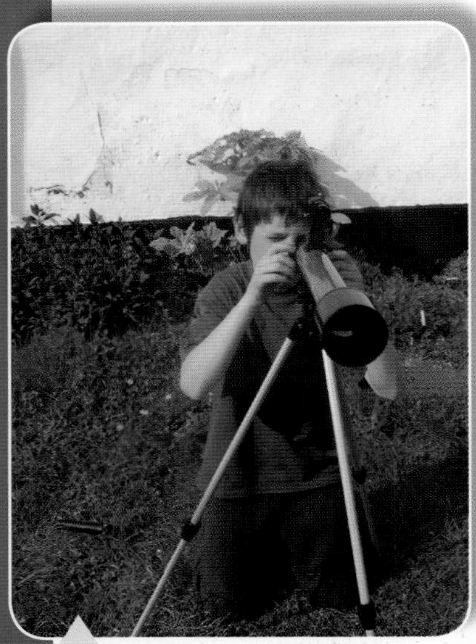

An amateur astronomer setting up his telescope – you can see the opening through which light enters the telescope

Seeing stars, seeing planets

When you look up at the night sky, light from the stars and planets enters your eye through the pupil. (Remember that the pupil is the black hole in the middle of your eye.) The pupil is small – perhaps 5 mm across – and it doesn't gather much light. That's one reason why astronomers use telescopes. An astronomical telescope has a 'hole' (the 'aperture') which may be 10 or 15 cm across. This lets in a lot more light, so we can see much dimmer objects.

Many telescopes make use of lenses. A **lens** is a specially shaped piece of glass which bends rays of light together to form an image.

Patterns in the sky

The stars form a fixed pattern in the night sky. The planets are different – they change their positions from one night to the next. They move slowly across the fixed background of the stars.

The planets change their positions because they are orbiting around the Sun – and so are we, the inhabitants of the Earth. All of the planets orbit the Sun in the same direction and in roughly the same plane. This means that, if you look at Venus on January 1st, it will be in one direction. If you look again on July 1st, it will be in a completely different direction.

The solar system

The solar system is made up of the Sun, all of the planets together with their moons and anything else which orbits the Sun (including asteroids and comets). There are plenty of other systems like this out in space – astronomers have detected planets orbiting around over 100 other stars.

The first four planets (Mercury, Venus, Earth and Mars) are the **rocky planets**. Beyond are the **gas giants**: Jupiter, Saturn, Uranus and Neptune.

a How many planets are there in the solar system?

Stretch Yourself

Many astronomical telescopes use curved mirrors to gather light. Find out how a curved mirror focusses light to produce an image.

Visiting a planet

Astronauts have visited the Moon; perhaps one day they will visit Mars. That would be a long and hazardous journey. It would take several years to go there and back.

It's much safer to send unmanned spacecraft to visit the other planets. We have a lot of information about the planets gathered by visiting spacecraft. Some fly straight past a planet, some go into orbit around it. Others land on the surface and send out robotic rovers to take pictures and even sample the rocks.

In this way, we can find out about the geology (rocks and internal structure) of the planet, its atmosphere, climate and magnetic field. The illustration gives an impression of the surface of one of Saturn's moons. Although Saturn itself is a ball of frozen gas, its moons are rocky and may have oceans of paraffin.

Next time you...

... look up at the stars in the night sky, picture the planets that orbit them.

A view from the surface of Titan, the largest of Saturn's moons

activity

Planetary explorer

One day, people may visit other planets in the solar system, or their moons. What conditions will they find when they get there?

- Choose a planet or a moon as your target. Find out what we already know about it.
- Now design a spacecraft and a robotic vehicle to visit the planet. How will your spacecraft gather information about the planet? It will need to have cameras on board, but what other instruments will it carry? How will you get the information back to Earth?

Summary Questions

1. Name the planets of the solar system in order, starting from closest to the Sun.
2. Besides the planets, what other objects are there in the solar system?
3. The illustration above shows an artist's impression of Saturn and Titan. Saturn is the sixth planet of the solar system. The Sun is shown shining brightly in the sky. Explain why the surface of Titan is likely to be much darker than this.
4. Astronauts have visited the Moon on several occasions. Some people suggested that these space missions were faked and that the photographs taken were made in a studio. Find out which features of the photos seemed suspect and the scientific explanations given by NASA for these features.

NASA astronauts on the Moon

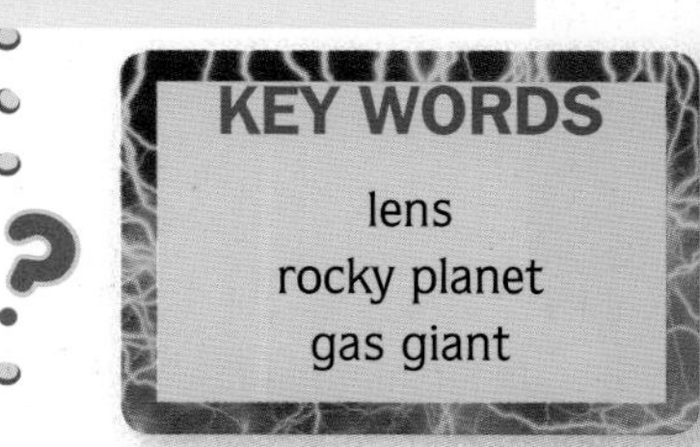

KEY WORDS

lens
rocky planet
gas giant

Phases of the Moon

- Why does the shape of the Moon seem to change?
- What does the Earth look like from the Moon?

These photos were taken on consecutive nights to show how the Moon's appearance changes

The changing Moon

The Moon looks pretty much the same as it did to your ancestors thousands of years ago. Very little happens on the Moon to change its surface.

However, we are used to the fact that the Moon's appearance changes from night to night. At full Moon, its bright disc rises in the East as the Sun sets in the West. At other times, it may appear as a half Moon, or just a thin crescent. These are the **phases** of the Moon. They go in order like this:

new Moon first quarter full Moon third quarter new Moon

a Look at the photographs which show the phases of the Moon. Which photo shows a full Moon?

We see the Moon because it reflects sunlight towards us. The Moon may look bright, but its rocky surface is really quite dark. It reflects only a small fraction of the light which falls on it.

At any time, just half of the Moon's surface is in sunlight. The other half is in shadow, just like the Earth.

Explaining the Moon's phases

Here is how to understand the phases of the Moon:

- As the diagram shows, the Moon orbits the Earth. It takes about 28 days to complete one orbit.
- The side of the Moon facing the Sun is brightly lit. The other side is in darkness.
- As the Moon travels round its orbit, the brightly lit part which we see changes.

To understand the diagram, first check that the side of the Moon facing the Sun is shown lit up. Then imagine looking from the Earth towards the Moon. What will you see?

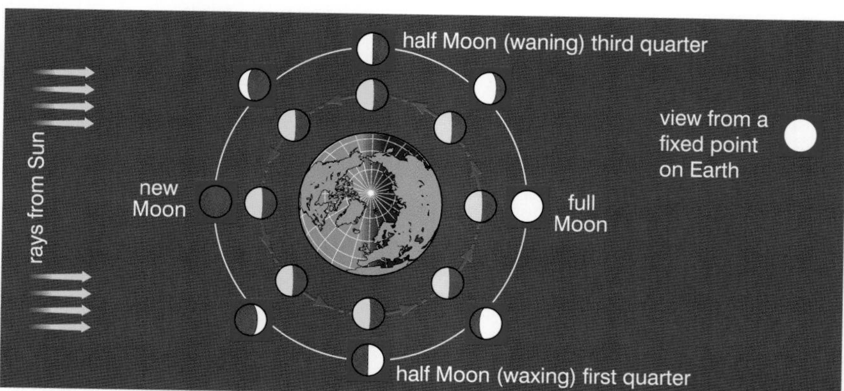

Understanding the Moon's phases

activity

Modelling the Moon

To model the phases of the Moon, you will need a lamp or other light source to represent the Sun. You will also need two spheres, a larger one to represent the Earth and a smaller one to represent the Moon.

Work out how to demonstrate the phases of the Moon using this arrangement.

Safety: Care – lamp may be hot. Take care if using mains electrical equipment.

The Moon from the Earth

It was in the early 1600s that scientists first used telescopes to study the Moon. They realised that its surface was covered in mountain ranges and deep **craters**. They imagined that the Moon's surface was similar to the Earth's and that the dark areas were oceans. Many suggested that there must also be plants, animals and even people living up there. They wondered what the inhabitants of the Moon would see if they looked at the Earth. Now we know, because we have photographs taken from the Moon's surface. However there is no-one there to enjoy the view!

Earth-rise, i.e. the Earth rising above the Moon's horizon

b Look at the photograph taken from the surface of the Moon. In which direction is the Sun?

activity

The Earth from the Moon

- What will the Earth look like, as seen from the Moon?
- Use the model you used earlier to investigate whether Moon dwellers will see 'phases of the Earth'.

We often see the Moon in the early evening. If it is full, it will be low on the horizon. If it is a first-quarter Moon, it will be high in the sky. Work out why this is so. Where will we see a new Moon?

Summary Questions

1. Draw a diagram to show how a ray of light from the Sun is reflected from the Moon and reaches our eyes here on Earth.
2. Draw diagrams to show the appearance of the Moon when it is full, at first quarter and at third quarter.
3. When the Moon is new, we cannot see it – its unlit side is facing the Earth. Look at the diagram on the opposite page, showing the phases of the Moon.
   a) Where does the new Moon appear in this series of photographs?
   b) How many days pass between full Moon and new Moon?
   c) How many days pass between one new Moon and the next?
   d) Find out what is meant by a 'gibbous Moon'.

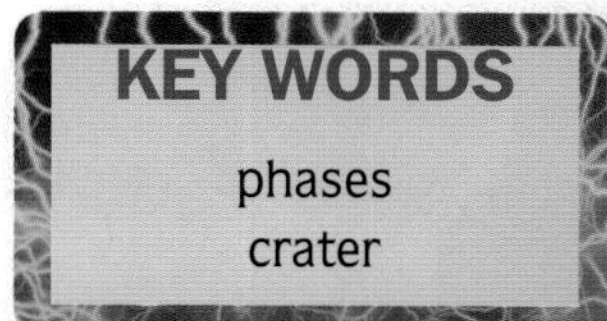

KEY WORDS

phases
crater

The Seasons

- How does the weather change during the year?
- Why does it change?
- Is it the same all round the world?

The yearly cycle

A year is the time it takes for the Earth to travel once around its orbit about the Sun – that's about 365¼ days. In the UK, we experience four **seasons** in the year – spring, summer, autumn and winter. Many of the living things around us have a life cycle which follows this pattern:

- Leaves appear on trees in the spring and fall off in autumn.
- Birds nest in the spring and may migrate to spend the winter in a warmer place.
- Children make snowmen in winter and have long holidays in the summer.

These things happen because we experience different weather patterns at different times of year. Days are longer and the Sun is higher in the sky in summer. Winter days are colder, darker and shorter.

There are other parts of the world where the pattern is very different. Tropical countries tend to experience two types of weather, the wet season and the dry season. Aboriginal people in northern Australia recognise seven seasons: lightning, thundering, rainmaking, greening, wind storming, fire raging and cloudless blue.

Seasons don't have definite dates when they start and end. They tend to merge into each other.

activity

Climate comparisons

The graph shows rainfall and sunshine figures for Bognor Regis. It is a seaside resort on the south coast of England and the UK's sunniest town.

Your task is to find similar data for other places around the world. You could look in travel brochures and websites and on the Met Office website. Find data for places in the northern and southern hemispheres and in the tropics (near the Equator).

- What patterns can you find in the data?
- Do they all show the same pattern as Bognor Regis?

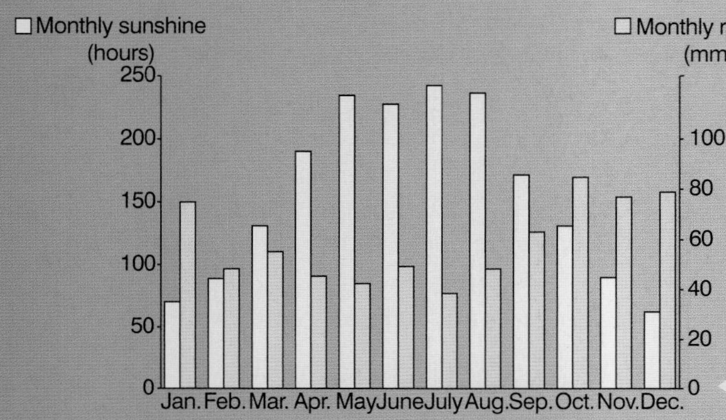

Rainfall (blue) and sunshine (yellow) data for Bognor Regis (average data, 1971–2000)

Explaining the seasons

As it travels round the Sun, the Earth spins on its **axis**, once each day. To understand why we have seasons, we need to know that the Earth's axis is tilted at about 23° to the plane of its orbit.

The diagram shows the Earth in two positions, at opposite ends of its orbit. Notice that the Earth's axis is tilted in the same direction in both positions. You should be able to see that:

- In December, the northern **hemisphere** is tilted away from the Sun. The southern hemisphere is tilted towards it.
- In June, the Earth is still tilted in the same direction. However now the northern hemisphere is tilted towards the Sun and the southern hemisphere is tilted away.

Because of this, in the UK we experience summer in June. The Sun rises earlier in the morning; at midday it is higher in the sky; and it sets later in the evening. With more hours of sunlight and with the Sun higher in the sky, we receive more radiation (light and heat) from the Sun. This makes our weather hotter.

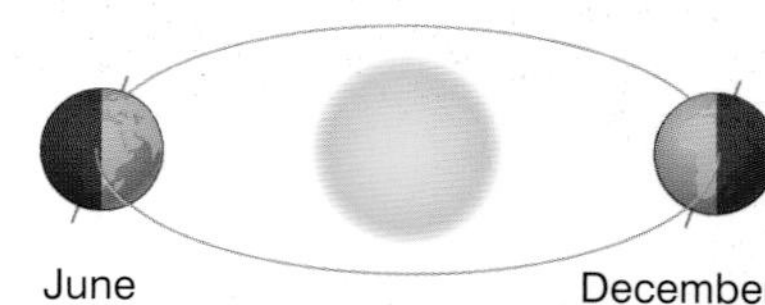

The Earth's axis is tilted relative to the plane of its orbit

a What can you say about the times of sunrise and sunset in winter, compared to summer? What can you say about the height of the Sun at midday on a winter's day?

When the Sun is higher in the sky, it feels much hotter. The diagram shows why this is. If the Sun is low in the sky, its rays are spread out over a larger area and so their heating effect is less.

activity

High in the sky

The Sun's rays provide us with light and heat. Their effect is greatest when the Sun is directly overhead.

- Devise a method to show that the Sun's rays have less effect when they strike the ground obliquely (at a low angle).
- Use a light meter to measure the brightness of the light.

Summary Questions

1. What do we mean by 'a year' and by 'a day'?
2. The diagram shows the Earth in June. Which part of the Earth is tilted towards the Sun at this time of year? Copy the diagram and add a line to represent the Earth's axis.

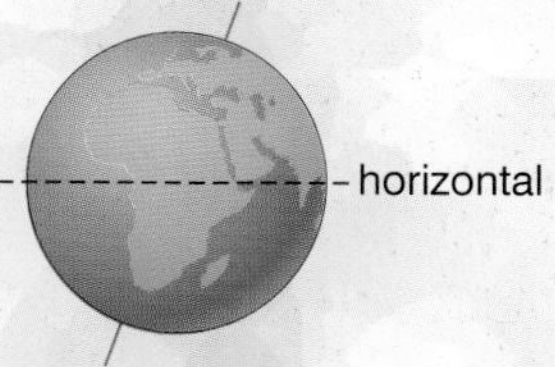

3. In June, the Sun is in the sky 24 hours a day at the North Pole. However, it never rises very high. Explain why this means that the summer at the pole is quite cold.
4. On the beaches near St Petersburg in Russia, people often sunbathe standing up. That way, they get a tan more quickly. Explain why. (St Petersburg is further north than anywhere in the UK.)

KEY WORDS

season
axis
hemisphere

Eclipses

- What happens during an eclipse?
- How can we explain eclipses?
- How can we use a model to explain observations?

When it all goes dark...

An eclipse of the Sun is a spectacular event, but rather rare. During a **solar eclipse**, for just a few minutes, the sky goes dark.

The Sun and Moon look the same size in the sky. A total eclipse happens because both are moving across the sky and once in a while the Moon's path takes it in front of the Sun.

It's pure chance that the Moon and Sun are the same size in the sky, but it does create a spectacular effect.

During an eclipse of the Sun, its corona (outer atmosphere) becomes visible

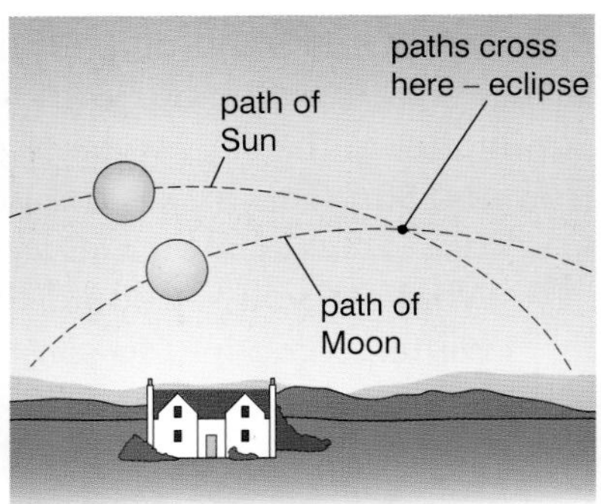

a The Moon is very slowly drifting away from the Earth. What effect will this have on solar eclipses?

Explaining a solar eclipse

For a solar eclipse to happen, the Sun, Moon and Earth must be exactly lined up. Then the Moon's shadow falls on the surface of the Earth.

Where the Moon completely blocks the Sun the shadow is total. This is called the **umbra**. At places close to this region, observers are in the **penumbra** and they will still see part of the face of the Sun. This is a partial eclipse.

Did You Know?

The last solar eclipse visible from the UK was in August 1999. It was cloudy, so few people saw it.

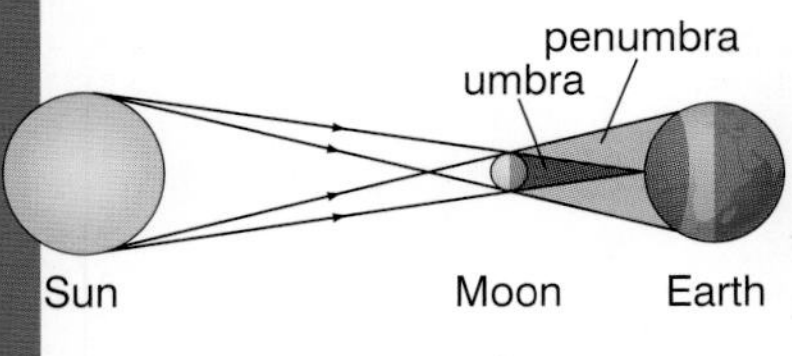

How a solar eclipse happens (not to scale)

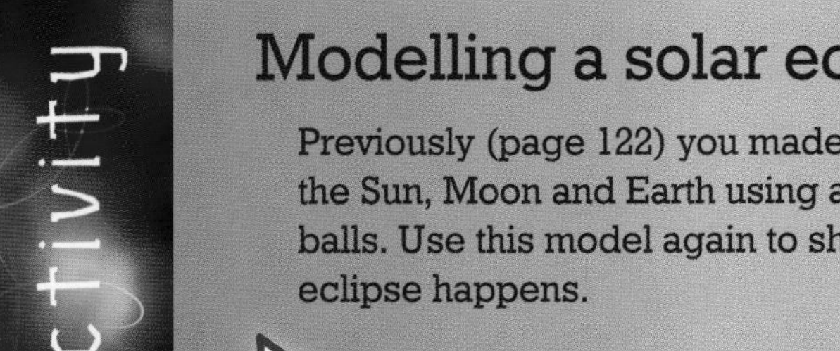

Modelling a solar eclipse

Previously (page 122) you made a model of the Sun, Moon and Earth using a lamp and two balls. Use this model again to show how a solar eclipse happens.

Safety: Care – lamp may be hot. Take care if using mains electrical equipment.

Eclipse of the Moon

Eclipses of the Moon happen more frequently than eclipses of the Sun. In a **lunar eclipse**, the Moon's orbit takes it through the Earth's shadow. The Moon becomes almost invisible. It is lit up as a dim red colour by sunlight refracted through the Earth's atmosphere.

From the diagram, you can see what happens in a lunar eclipse. The Sun, Earth and Moon are lined up with the Moon on the far side of the Earth. Anyone on the dark (night) side of the Earth will see the eclipse.

The Moon, photographed as it goes through an eclipse

activity

Modelling a lunar eclipse

- Use the lamp-and-balls model to show how a lunar eclipse happens.

Safety: Care – lamp may be hot. Take care if using mains electrical equipment.

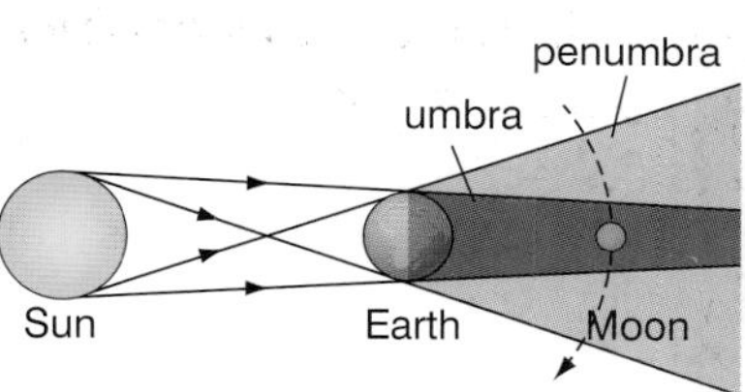

How a lunar eclipse happens (not to scale)

1. What is eclipsed (hidden) during a solar eclipse? What eclipses it?
2. What three objects are involved in a lunar eclipse? Draw a diagram to show how they are arranged during a lunar eclipse.
3. The Earth is much bigger than the Moon (about four times the diameter). Explain why this makes a lunar eclipse more likely than a solar eclipse.
4. Look at the diagram above showing a lunar eclipse. What phase must the Moon be at for this to happen? What will its phase be at the time of a solar eclipse?
5. The Moon is very, very gradually drifting away from the Earth.
   a) What effect will this have on the size of the Moon in the sky as seen from Earth?
   b) What effect will it have on solar eclipses?

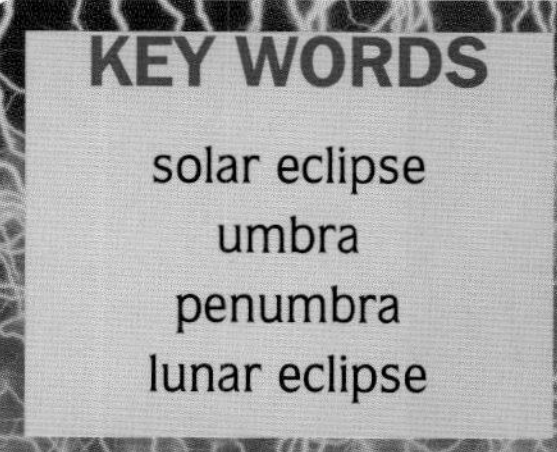

KEY WORDS
solar eclipse
umbra
penumbra
lunar eclipse

know your stuff

▼ Question 1 (level 4)

The diagram shows a ray of light shining on to a flat mirror.

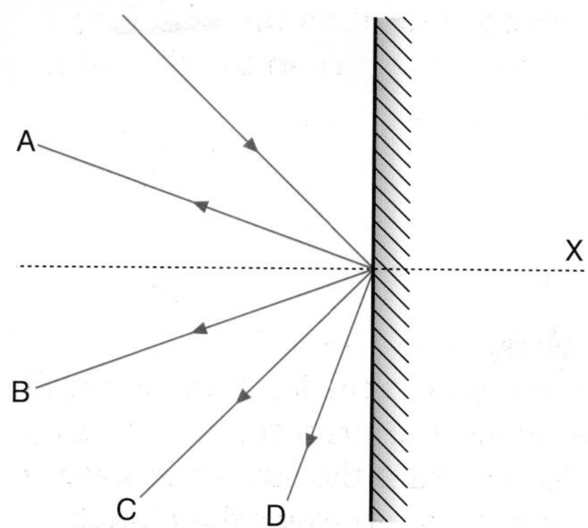

a Which of the rays (A, B, C or D) shows correctly the path of the ray after it has been reflected by the mirror? [1]

b Name the line marked X. [1]

c Copy the diagram, showing the incident and reflected rays. Label these rays. [2]

d Mark two angles which are equal. [1]

▼ Question 2 (level 5)

Jenny is experimenting with a glass block. She shines a narrow ray of light at it. She tries different positions of the glass block.

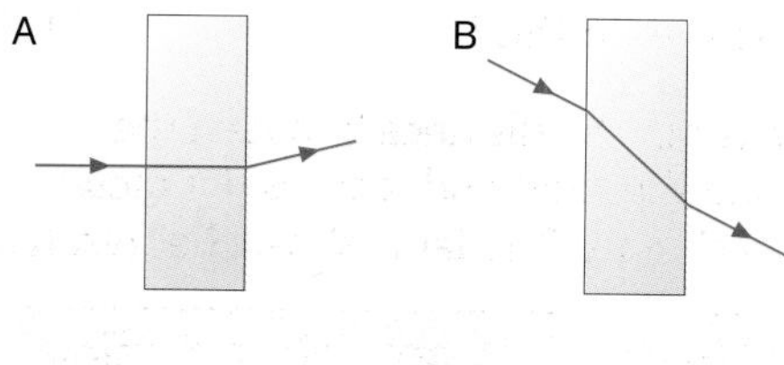

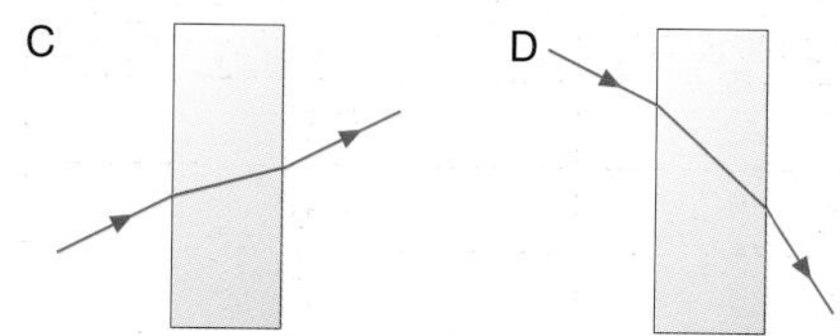

For each of the diagrams A to D, say whether it shows correctly how the ray could have behaved. [4]

▼ Question 3 (level 5)

The drawing shows the Earth with the Sun's rays shining on it. Give reasons to support your answer to each of the questions that follow.

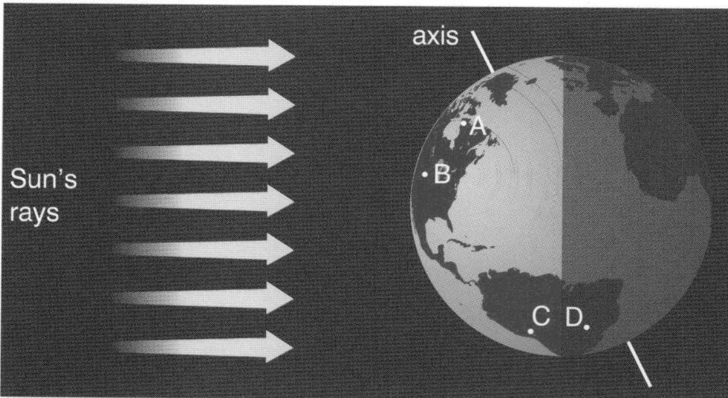

a At which point is it night-time? Why? [2]

b At which point is the Sun highest in the sky? Why? [2]

c At which points is it winter? Why? [3]

▼ Question 4 (level 5)

The diagram shows the Sun, Earth and Moon (not to scale). The Sun is a luminous object.

a What do we mean by a 'luminous object'? [1]

b Name another luminous object in space. [1]

c Copy the diagram. Add rays to show how light from the Sun allows people on Earth to see the Moon. [2]

d A few days later, people on Earth observed an eclipse of the Moon. On your diagram, show the position of the Moon during the eclipse. [1]

e Mark with a cross (×) a point on the Earth's surface from which the eclipse would be visible. [1]

How Science Works

▼ Question 1 (level 6)

Jake was investigating how light is refracted as it passes through a glass block. He used a ray box to produce a narrow ray of light.

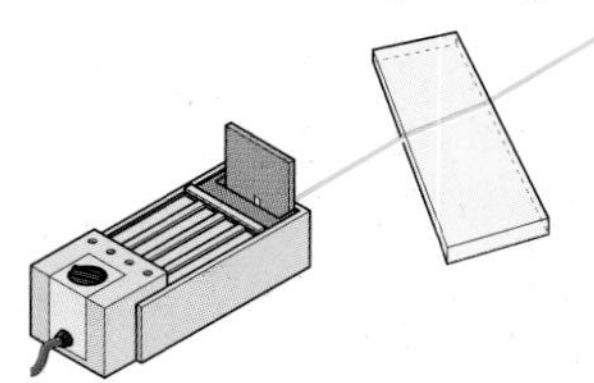

Jake recorded the path of the ray and the position of the glass so that he could draw a diagram and measure the angles of incidence and refraction.

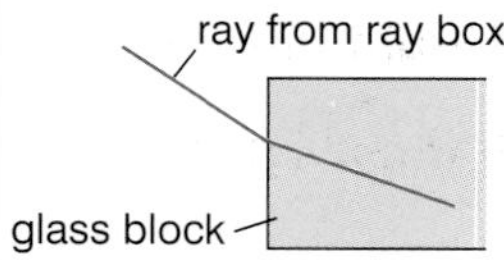

a Copy the diagram of the glass block and ray of light; mark and label these angles:

(i) angle of incidence

(ii) angle of refraction. [2]

Jake did this experiment three times. The table shows his results.

Angle of incidence	Angle of refraction
32°	21°
38°	24°
44°	28°

b Which *two* of the following are reasonable conclusions to draw from Jake's results? Give the letters.

A The angle of incidence is greater than the angle of refraction.

B The angle of refraction is greater than the angle of incidence.

C The ray always bends away from the normal when it enters glass.

D The ray always bends towards the normal when it enters glass. [2]

c Jake's experiment would have been better if he had used a greater *range* of values for the angle of incidence. Give *two* additional values of the angle of incidence, each of which would have increased the range. [2]

▼ Question 2 (level 7)

Astronomers have detected a planet orbiting around a distant star. Its orbit has a different shape from the Earth's orbit around the Sun; it is an elongated ellipse, as shown in the drawing.

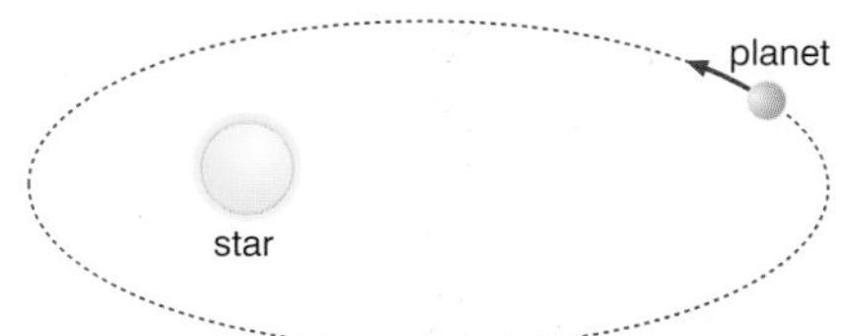

Janet is investigating how the light falling on the planet will change as it orbits a star. She sets up a bright lamp to represent the star and moves a light meter around it to represent the planet.

a Why is it important that Janet works in a darkened room? [1]

b In each position of the 'planet' around its orbit, Janet adjusts the light meter, tilting it around until it gives the maximum reading. Why does she do this? [1]

Janet measures the distance between the 'planet' and the 'star' and records the meter reading each time. The table shows her results.

Distance between star and planet (m)	Light meter reading
1.20	0.6
0.90	1.6
0.60	3.8
0.30	14.9

c Draw a graph to represent Janet's results. [2]

d Janet felt that she might have made a mistake in the final reading (light meter = 14.9) and that it did not fit the pattern of the other results. Suggest *two* things she could do to check this idea. [2]

Heat and Sound

P2.1

Yes, wool is warm

Woolly ideas

Sheep can survive outdoors in cold weather. Sheep have been bred for thousands of years to have long, woolly coats. Every year, they are shorn to provide us with the wool we use to make warm clothing. Then we, too, can go out in the snow.

Humans and sheep are mammals. Mammals, including humans, have a body temperature of about 37°C. That's a lot higher than the temperature on a snowy day. We need to wear thick clothing to stop us feeling cold. If you have taken part in winter sports such as skiing and snowboarding, you have probably worn specially designed clothing. This keeps you warm and at the same time allows you to move freely.

a Look at the photo of the skier, flying through the air. How are his clothes designed to keep him warm? What fabric might they be made of? Why is the snowman wearing a woolly hat and scarf?

activity

Feeling warm

We think of wool as a 'warm' material. It feels warm when we touch it. Our skin has nerve endings which tell us if the materials we are touching are hot or cold.

Handle some different materials. Try the following:

- Fabrics such as wool, nylon, cotton
- Solid materials such as wood, steel, aluminium, polystyrene, polythene
- A beaker containing water.
- Put your samples in order, from 'coldest' to 'warmest'.
- Does your order agree with the rest of the class?
- How easy is it to judge whether one material is warmer than another?
- Can you identify any factors which affect how warm the material feels?

Seeing heat

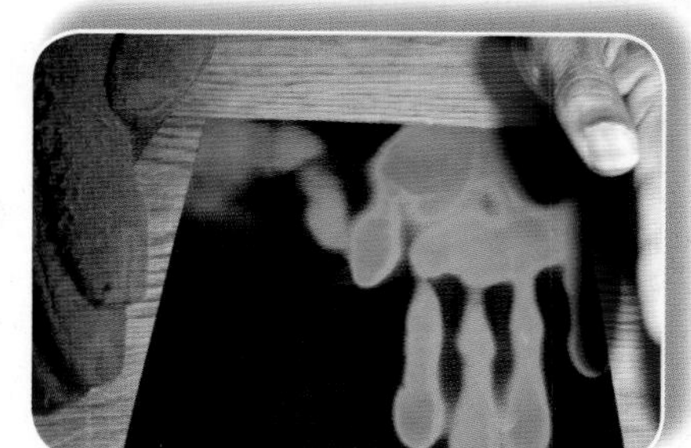

Thermocolour film is a special kind of plastic. If you warm it, it changes colour. You can use it to see where the temperature is higher and where it is lower. Compared with touching things by hand, it's a more scientific and reliable method of telling how warm something is.

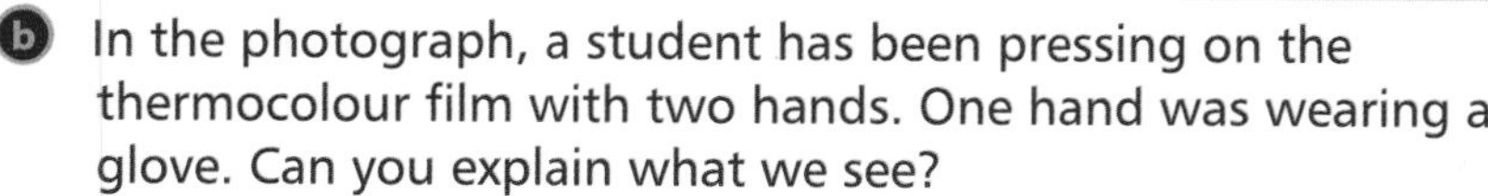

b In the photograph, a student has been pressing on the thermocolour film with two hands. One hand was wearing a glove. Can you explain what we see?

activity

Colourful heat

Design some experiments using thermocolour film. For example, you could place a hot cup of water on a thick book. How could you show that heat from the cup has travelled into the book?

How could you use thermocolour film to compare the materials you investigated by touch in the experiment on the opposite page?

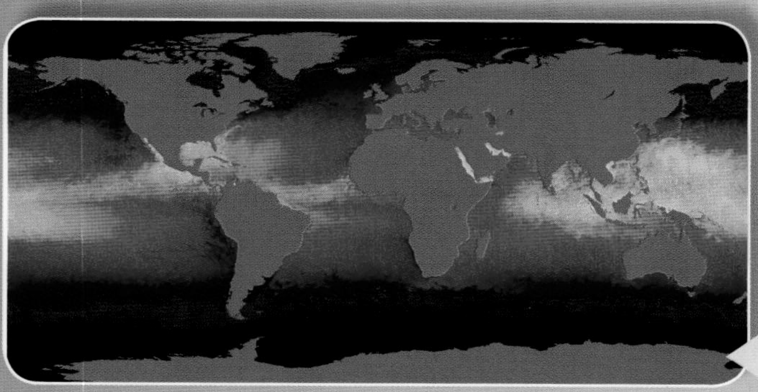

This map shows where the oceans are hottest (yellow) and coldest (black)

Summary Questions

You can have my wool if I can have your coat.

1. Explain why thermocolour film is a more scientific way of testing 'warmness' than using your fingers.
2. Hold some everyday objects gently against your top lip. (Try a plastic pen, a wooden pencil, a china cup, a metal teaspoon.) Your lip is very sensitive to heat. Explain what we mean by 'sensitive'. Why do you think it is important that your lip is sensitive to heat?
3. Look at the map which shows how the temperature of the Earth's oceans varies. The data were collected by a satellite in orbit around the Earth.
   a) What patterns can you see in the map?
   b) Many people are concerned about climate change. Why might maps like this be useful?

Taking Temperatures

- How can we measure temperatures accurately?
- Are some thermometers better than others?

Feeling is believing?

In everyday life, we rely on the nerve endings in our skin to tell us about how hot things are around us. We touch things to see if they are hot or cold. We don't even have to touch. If you hold your hand close to an object, you may be able to feel heat coming from it. However our senses can deceive us – that's why scientists often rely on instruments to make measurements.

- **Temperature** tells us how hot or cold something is.
- The instrument used to measure temperature is the **thermometer**.

The next experiment shows why it is better to use a thermometer to measure temperature than to rely on what we feel.

activity

Getting warmer

For this experiment, you need three bowls of water. One should be warm, one cold and one a mixture of warm and cold, so that its temperature is between the other two.

- Place your left hand in the warm water and your right hand in the cold water. You will feel the difference. Keep them there for a minute or so.
- Now place both hands in the third bowl. What do you notice?
- Use a thermometer to check the temperatures of the water in the three bowls.

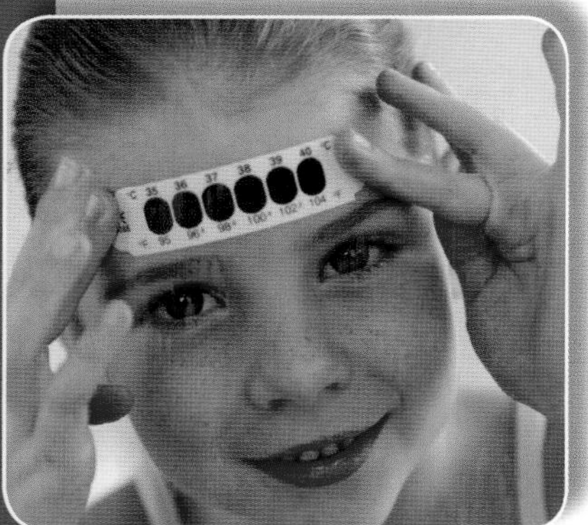

Unreliable evidence

In the experiment, the hand that has been in cold water feels that the third bowl is warm. The other hand feels that it is cold. Of course, they should both feel that the water is at the same temperature.

If you are unwell, someone may touch your forehead to see whether you are hotter than normal. A doctor or nurse will use a thermometer to measure your temperature on the **Celsius scale**. The thermometer may be a strip of thermocolour film. The strip has sections which 'light up' to show your temperature. A thermometer like this has only a small **range** – from 35°C to 40°C, perhaps.

a When a parent is about to bath a small baby, they may dip their elbow in the bath water first, before putting the baby in. Why do they do this?

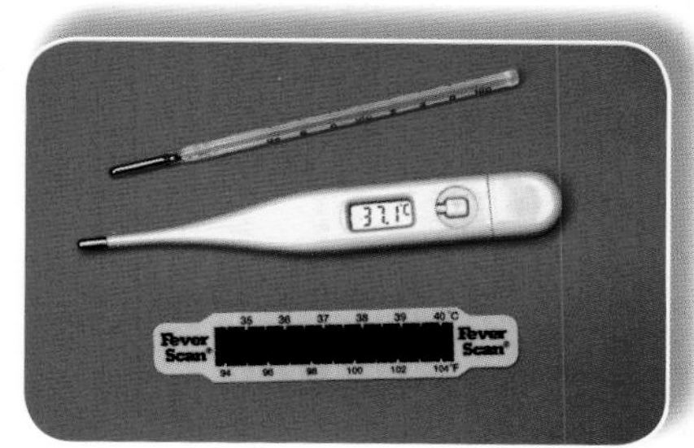

Types of thermometer

There are many different types of thermometer. Here are a few commonly used in the lab.

- **Liquid-in-glass:** Alcohol or mercury expands as it gets hotter and rises up the tube.
- **Electronic:** The sensor's electrical resistance changes and the display shows the temperature.
- **Liquid-crystal:** Different colours show up as the temperature changes.

A matter of degree

It is useful to know some temperatures on the Celsius scale. For example, comfortable room temperature is about 20°C. The picture shows some more.

Did You Know?

The lowest temperature ever recorded on Earth was –91°C at Vostok in Antarctica in 1997.

activity

Looking at thermometers

- Examine some thermometers. Which types are they?
- Use them to measure some temperatures. Try water fresh from the tap and water which has been standing in the room for some time.
- Now compare the thermometers. Think about:
  - Can you tell how **accurate** they are? How close do they get to giving the correct temperature?
  - How **sensitive** are they? Can they tell you if the temperature changes by, say, 1°C or even 0.1°C?
  - What is their **range**? Find out the lowest and highest temperatures they can measure.

Summary Questions

1. At what temperature does water freeze? At what temperature does it boil?
2. a) Explain what we mean by the 'range' of a thermometer.
   b) Doctors use a clinical thermometer to measure a patient's temperature. Why does this type of thermometer have a small range?
3. Look at the photographs of the three thermometers at the top of this page. Which is the most sensitive? Explain how you know.
4. We have only looked at three types of thermometer. Find out about some more and say what they are used for.

KEY WORDS

temperature
thermometer
Celsius scale
range
accurate
sensitive

Warming up, Cooling down

- Why do things warm up or cool down?
- What is the difference between 'heat' and 'hotness'?

Cooling off

A cup of tea may be too hot to drink. Wait a while and it will cool down to a comfortable temperature. Why is that?

The drink cools down because it is losing **heat energy** to its surroundings. The diagram shows what's going on.

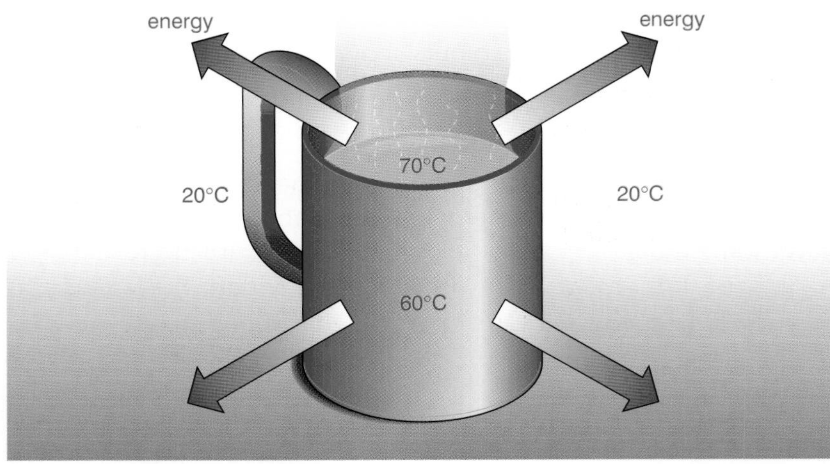

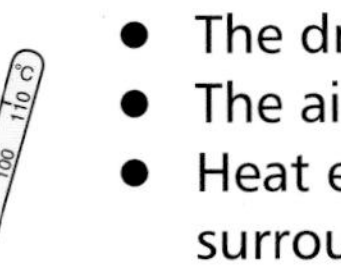

- The drink is at a temperature of 70°C.
- The air and table are at 20°C.
- Heat energy spreads out from the hotter drink to its cooler surroundings.

Heat energy moves from a warmer object to a colder one. It is the *difference in temperature* that makes the heat energy move. As an object loses heat energy, its temperature drops.

a If the drink in the diagram was hotter – say, 100°C – the temperature difference would be greater. Make a prediction: would heat energy leave the drink more quickly or more slowly?

activity

Patterns of cooling

- Find out the pattern of cooling. Start with a beaker of hot water.
- Use a thermometer to measure its temperature. Record the temperature every minute as the water cools. Then draw a graph. Alternatively, use a temperature sensor. Connect it to a data-logger which will record the data for you. Then a computer will draw the graph for you.
- What pattern do you see? Can you explain it?

The importance of temperature difference

At first, the temperature of the water drops rapidly. It is much hotter than its surroundings (there is a big temperature difference), so heat energy leaves quickly.

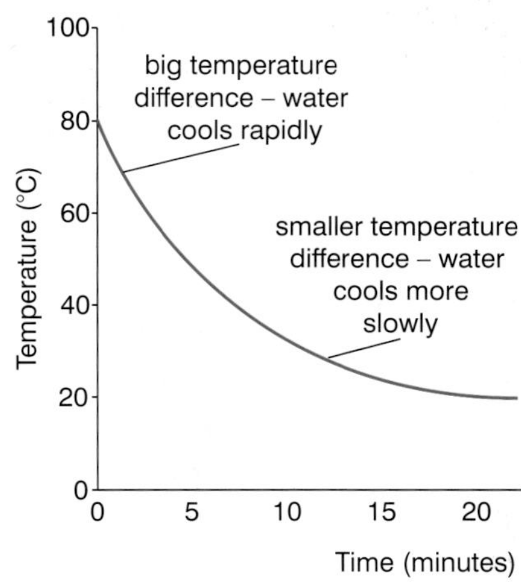

As the water gets colder, the difference in temperature between the water and its surroundings is less. This means that heat energy leaves more slowly, so the temperature of the water drops more slowly.

Our body temperature is about 37°C. In the UK, the outside temperature rarely gets above 30°C and in winter it is often below 0°C. This means that our bodies are usually warmer than our surroundings and so we lose energy all the time. How do we cope with this?

- We live in buildings which are often heated – so that the temperature difference between us and our surroundings is small.
- We wear clothes (thicker ones in winter) – to try to prevent heat from escaping.

The photograph shows a thermogram of a fridge. The colours tell you the temperatures, from purple and blue (lowest temperatures) to red (highest). Imagine putting a warm object from the room into the fridge. There is a temperature difference so that heat energy flows out of the object and its temperature falls.

A thermogram uses colours to show temperatures

Heat and hotness

Temperature is a measure of an object's 'hotness' (how hot it is).

Heat energy (or simply heat) is energy moving from a hotter place to a colder place, because of the difference in temperature.

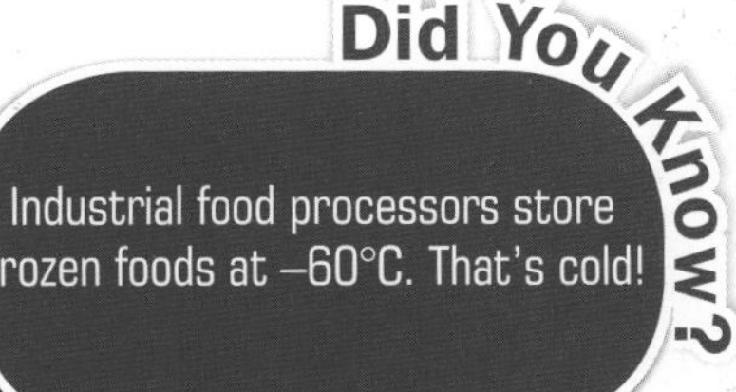

Did You Know?
Industrial food processors store frozen foods at –60°C. That's cold!

Summary Questions

1. a) What one word means 'energy moving from a hotter place to a colder place'?

   b) What one word means 'how hot something is'?

2. Look at the thermogram of the fridge.
   a) Which shelf is the coldest? How can you tell?
   b) Name an object which has recently been put in the fridge. How can you tell?

3. Look around the room that you are in. Which things are warmer than room temperature? Which are colder? How does heat flow in the room? Draw a diagram of the room and use arrows to show how heat is flowing.

Next time you...

... have a hot drink, picture the heat energy spreading out from it. Think about the temperature difference which is making the heat spread out.

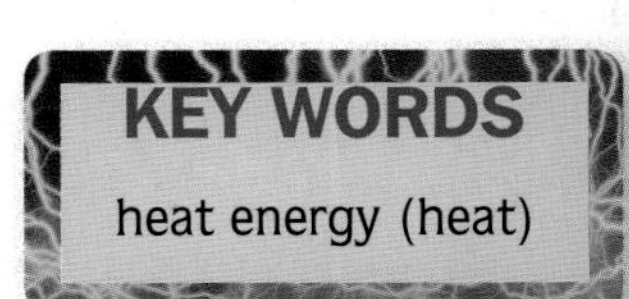

Thermal Conduction and Insulation

- Which materials conduct heat well?
- Which are good insulators?

These young people would soon feel cold if they weren't wearing insulating clothing

science@work

Scientists are developing smart clothing. The fabric changes so that it is warm on a cold day but cool on a hot day.

Slowing the flow

In cold weather, we want to slow down the rate at which heat escapes from our bodies. That's why we wear thick clothes.

Clothes for cold weather are made of materials like wool. Wool is a good **thermal insulator** – heat only travels slowly through wool. If you wear a woollen jacket on a cold day, the temperature on the inside will be about 37°C – body temperature. The temperature on the outside might be 0°C, so the difference is 37°C. That's a big difference, but heat leaves your body only slowly because wool is a good insulator. Take off your clothes and you will soon notice the difference!

Too hot to handle

Put a metal teaspoon in a hot drink. Very soon, the end of the spoon will be hot. Heat energy has travelled up through the spoon from the drink.

Metals are good **thermal conductors** (conductors of heat). They allow heat to pass through them easily.

activity

Comparing metals

In the diagram, the two rods are made of different metals, copper and steel. They are the same length and diameter.

Each rod has a temperature sensor at one end and is heated at the other end.

A data-logger and computer show how the temperature of the end of the rod rises as the other end is heated.

- Which variables should we control to make this a fair test?
- Which metal is the better conductor of heat?

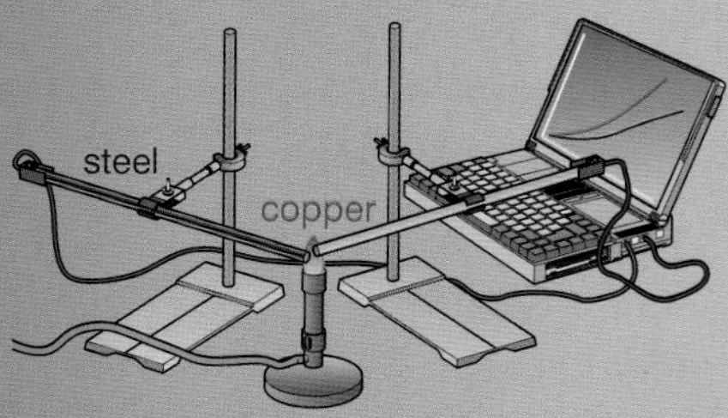

Safety: Metal objects will stay hot for some time after heating has finished.

a Which features of this experiment are designed to make it a fair test?

Conductors and insulators

Metals are good conductors of heat, but some are better than others. Copper is twice as good as aluminium. Steel is a poorer conductor than aluminium.

Non-metals are mostly insulators of heat. Plastics, wood, glass, china – all of these are insulators. So are materials which contain a lot of air, such as expanded polystyrene, wool and cotton. (When you wear woollen clothes, it is really the air in between the woollen fibres which is insulating you.)

Gases, such as air, are very good insulators.

Deceptive feelings

A metal ruler feels cold when you touch it; a plastic ruler feels warm. In fact, they are both at room temperature. How do our senses deceive us like this?

When you touch a metal ruler, heat from your warm finger flows into the metal. This cools your skin and your nerve endings send a signal to your brain saying that it's cold.

Heat cannot flow into a plastic ruler because plastic is an insulator. Your finger stays warm.

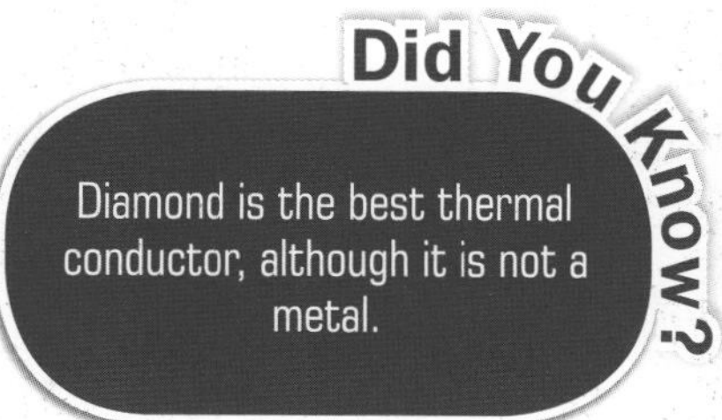

Did You Know?

Diamond is the best thermal conductor, although it is not a metal.

b Look back at the experiment 'Feeling warm' on page 130, where you investigated 'warm' and 'cool' materials. Can you make sense of your list? When we say that a material is 'warm', what do we mean?

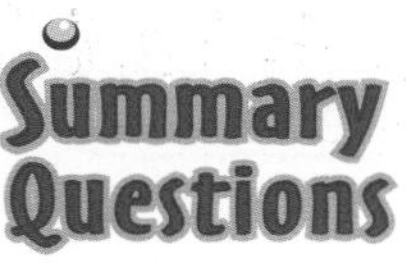

Summary Questions

1. Name three materials that are good conductors of heat and three that are good insulators.
2. Which of the following materials is the best conductor? Which is the best insulator?

   aluminium copper steel wood glass air
3. Design an experiment to compare two materials, to find out which is the best insulator. Start with a beaker of hot water; how does its temperature change when it is surrounded by insulation? How will you make your experiment a fair test? How will you tell from your results which material is the better insulator?
4. If you sleep under a duvet, you may have to change from one duvet to another when winter comes along. Find out about the tog ratings which tell you about the insulating properties of different duvets.

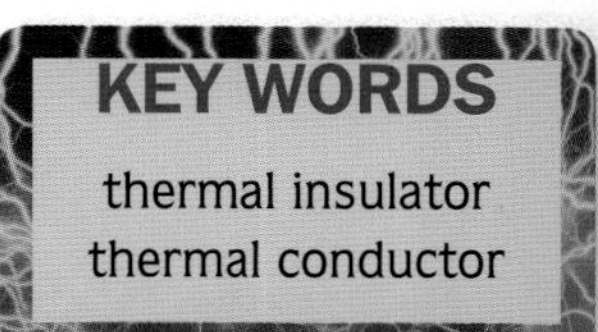

KEY WORDS

thermal insulator
thermal conductor

Expansion and Contraction

- Why do solids, liquids and gases expand when they are heated?
- How does heat travel through solids?

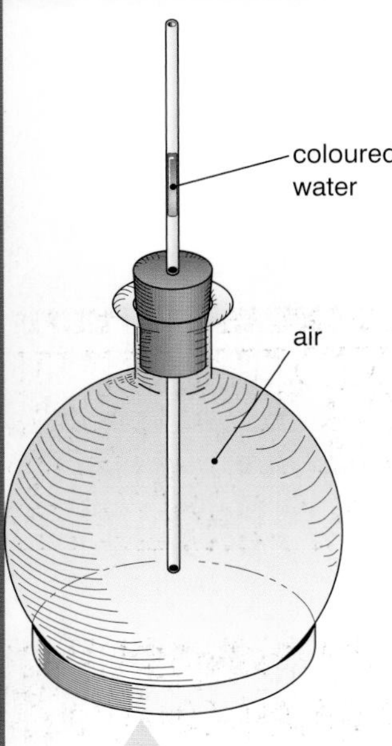

A model of Archimedes' thermometer

Next time you...

... cross a motorway bridge, look for the expansion joints in the roadway. Bridges get longer on a hot day and need these joints to allow for this.

The heat is on

Over 2000 years ago, the Greek scientist Archimedes invented a thermometer. You can see a modern version of it in the diagram.

On a hot day, the air in the flask expanded and pushed the water up the tube. When the temperature dropped, the air contracted and the water moved back down the tube.

Archimedes had noticed that air expands when it is heated. Because it only expands a little, he used a big flask and a narrow tube. That way the water moved a long way for a small change in temperature.

a Look at Archimedes' thermometer. Explain why there must be no leaks around the stopper. How could the design be changed so that the thermometer works over a bigger range of temperatures?

Materials expanding

Most materials **expand** when they get hotter and **contract** when they cool down. Solids, liquids and gases all behave like this.

Sometimes, expansion can be a problem. On very hot days, railway lines may expand and buckle. Then the trains stop running and engineers have to repair the tracks. Electricity cables contract in cold weather. If they contract a lot, they may snap.

activity

Observing expansion and contraction

Watch some demonstrations which show how materials expand when they are heated and contract when they are cooled.

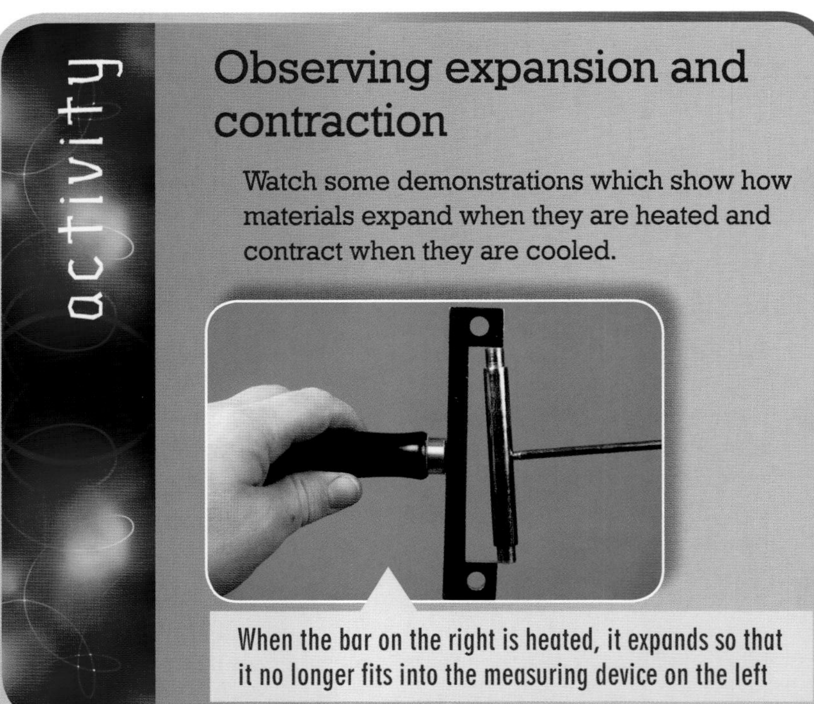

When the bar on the right is heated, it expands so that it no longer fits into the measuring device on the left

Particle explanations

We can use the **particle model** of matter to explain why materials expand when they are heated.

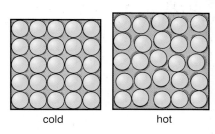

- Solid materials are made of particles. These particles are packed closely together. When the material is cold, the particles vibrate a little about their fixed positions.
- When the material is heated, the particles are given energy so they begin to vibrate more. Each particle pushes on its neighbours, increasing the average distance between them. So the solid gets bigger – it expands.

cold hot

Notice that the particles themselves don't get any bigger – they just take up more space.

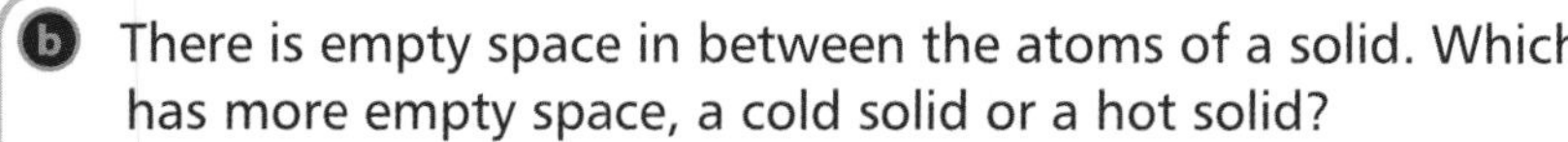

b There is empty space in between the atoms of a solid. Which has more empty space, a cold solid or a hot solid?

Explaining conduction

How does heat energy travel through a solid material? Here is one way; we will use the particle model again:

- The diagram shows the particles in a long rod which is being heated at one end. Heating the rod gives energy to the particles, so that they vibrate more and more.
- The particles bump against their neighbours, sharing their energy with them.
- Gradually, energy is shared along the rod. Eventually, all of the particles in the rod have gained energy and are vibrating more vigorously.

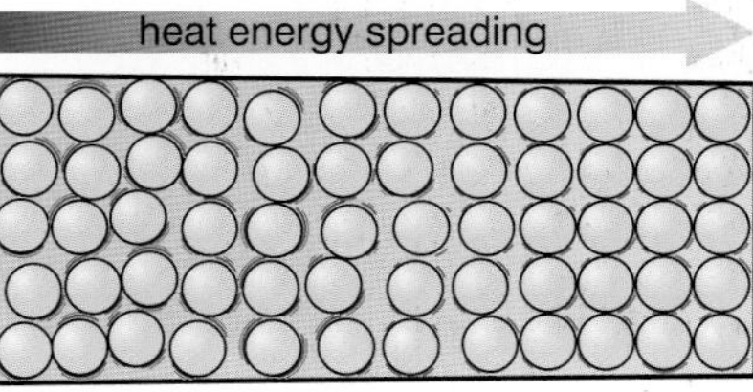

In a gas, the particles are far apart, so they rarely bump into each other. This means that energy doesn't spread so rapidly through a gas.

Help Yourself

You have studied the particle model of matter in Chemistry lessons. Look back to recall how ideas about particles can explain some of the differences between solids, liquids and gases.

Summary Questions

1. Copy these sentences and replace the words in italics with the correct scientific words.

   When a steel bar is heated it *gets bigger*.
   When it cools down, it *shrinks*.

2. You have measured temperatures using a liquid-in-glass thermometer. Explain how this makes use of expansion.

3. How do the particles in a solid material move as it gets hotter?

4. You have learned that the particles in a liquid are close together, as in a solid, but more randomly arranged. Write a few illustrated sentences to explain why a liquid expands when it is heated.

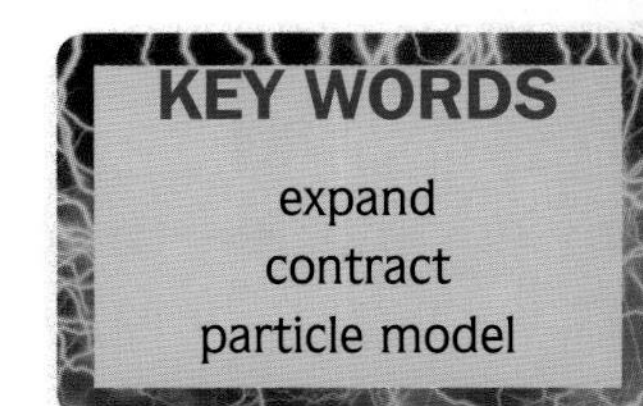

Radiation and Convection

P2.6

- How does heat energy reach us from the Sun?
- How does heat energy spread around a house?
- Why does hot air rise?

Heat spreading out

Any warm or hot object gives out heat to its surroundings. This is called **radiation** and it is different from conduction.

- In conduction, heat travels *through* a material. The particles of the material collide with each other, sharing their energy.
- In radiation, heat travels outwards in all directions in the form of **infra-red** radiation.

Heat can radiate through empty space – that's how energy from the Sun reaches us. Heat can also radiate through a transparent material like air. Hold the back of your hand close to a cup of hot drink. You may be able to feel heat radiation coming from the cup.

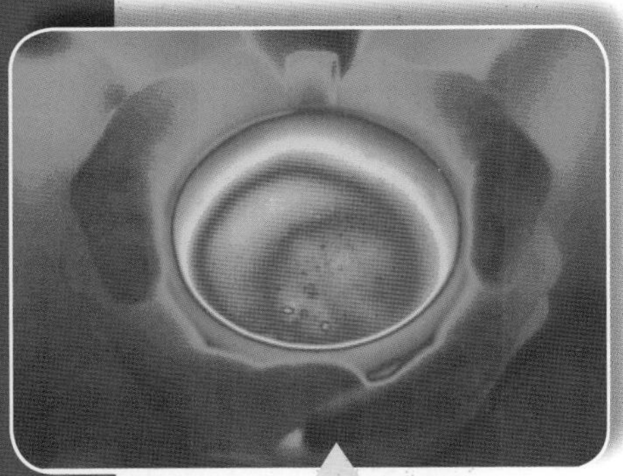

This image was made using infra-red radiation. You can see that the drink is hottest – it is giving out lots of radiation

a Look at the infra-red image of the person holding a cup of hot coffee. The drink is giving out lots of infra-red radiation. Why is the top of the cup less bright? You can see that the person is also giving out infra-red radiation, but less than the drink. Which parts of their hands are the coolest? How can you tell?

activity

Seeing infra-red radiation

A video camera can be adapted to show up infra-red radiation. Use one to look at some objects, hot and cold, in the lab.

Wildlife photographers use infra-red radiation to see what is going on in the dark

Air movement at a bonfire

Heat moving upwards

If you light a bonfire, flames, smoke and sparks rise upwards, carried by hot air. Cold air flows in to replace the hot air and, in turn, is heated by the fire. You have created a **convection current**.

activity

A model convection current

You can have convection currents in liquids (such as water) as well as in gases (like air).

- Use crystals to colour the water so that you can see how the current flows.

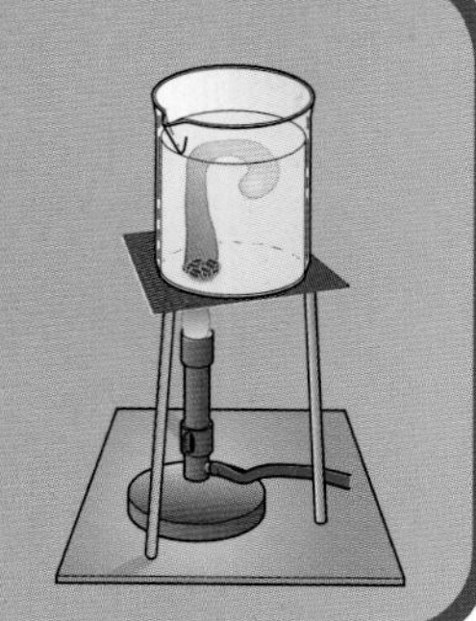

Safety: Use forceps to handle the coloured crystals.

Explaining convection

Why does hot air rise like this? We can explain convection using the particle model and ideas about forces.

- When air is heated, its particles get more energy. They push each other farther apart. This means that the **density** of the air is less.
- Because the warm air is less dense than the surrounding air, it floats upwards. The **upthrust** on it is greater than its weight.
- Colder air moves in to replace it.

Convection currents are very important in our lives. The wind is a convection current and so are ocean currents. Understanding these currents allows us to understand our climate and predict the weather.

Summary Questions

1. What processes are being described here?
   a) Heat energy travelling along a metal rod.
   b) Heat energy spreading out from a hot object through a vacuum.
   c) Heat energy carried by a current of water or air.
2. Hot air balloonists must understand convection currents. Explain why.
3. Some houses have open fires in which coal or wood is burnt. Explain with a diagram how a convection current ensures that the smoke goes up the chimney.
4. During the day, radiation from the Sun warms up our surroundings. Use the idea of radiation to explain why it gets cold at night and why it is coldest just before dawn.

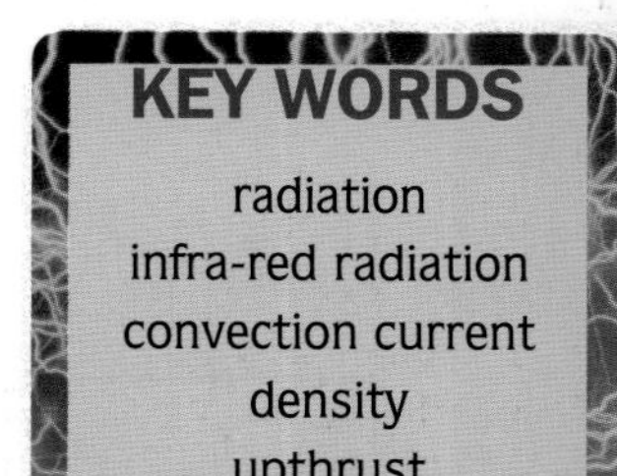

KEY WORDS

radiation
infra-red radiation
convection current
density
upthrust

Seeing Sounds

- How does a sound wave change when the loudness changes?
- How does a sound wave change when the pitch changes?

Getting louder

Sounds are produced when something vibrates. A guitar has strings which vibrate. In a wind instrument like a trumpet, it is the air inside which vibrates when you play a note.

To make a louder note, you have to make the guitar strings vibrate more, or blow harder into the trumpet. The sound is louder, but you can't really see why this is. When a loud sound reaches your ears, how does it differ from a soft sound?

You can investigate this using three pieces of equipment:

- A **signal generator** makes sounds which can be loud or soft. You can also change the **pitch** (how high or low the sound is).
- A **loudspeaker** lets you hear the sound from the signal generator.
- An **oscilloscope** shows a trace of the sound on its screen.

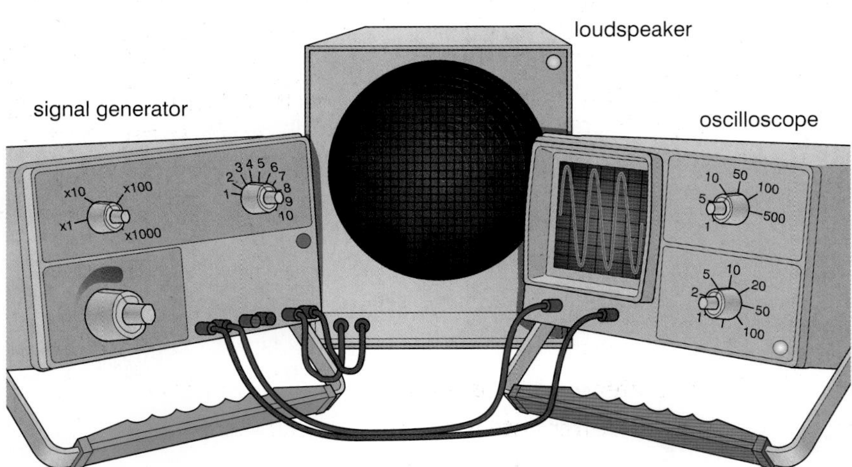

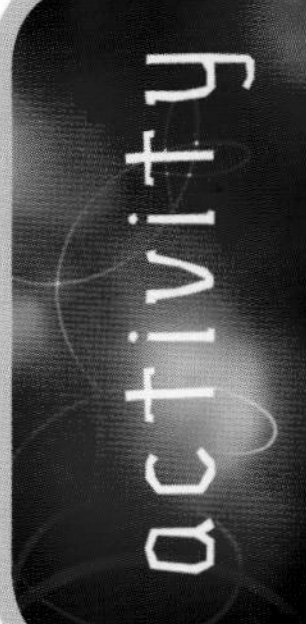

Sounds on screen

- Connect the equipment.
- Start by watching how the trace changes when a sound gets louder and quieter.
- Then watch it change as the pitch of the note changes. What patterns can you see?

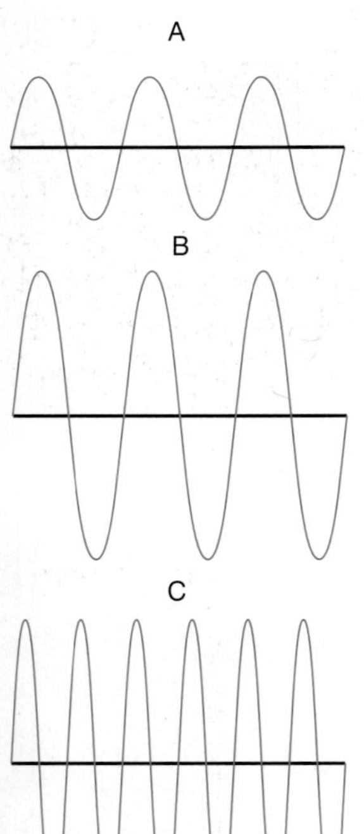

Traces of sound

The oscilloscope trace is made by a dot travelling across the screen. It goes up and down to show the vibrations of the sound.

The diagrams show how the trace changes as the sound changes.

As a sound gets louder, the trace goes up and down more. Its **amplitude** has increased. (Amplitude is the height of the trace above the centre line.)

As a sound gets higher, the trace goes up and down more rapidly. Its **frequency** has increased. (Frequency is the number of vibrations each second.)

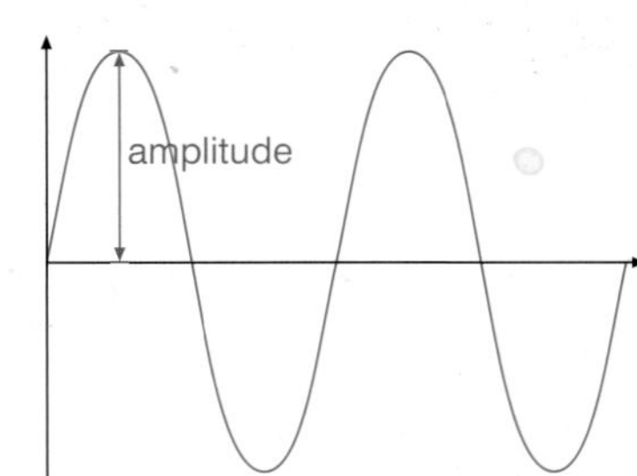

a Look at traces A, B and C on the opposite page. Which two have the same loudness? How can you tell?

b Which trace has the highest frequency?

activity

Capturing sounds

- Connect a microphone to a data-logger – this will record sounds.
- Connect the data-logger to a computer – this will show the sounds you have recorded.
- Play some different musical instruments next to the microphone.
- Can you see any difference in their traces on the screen?

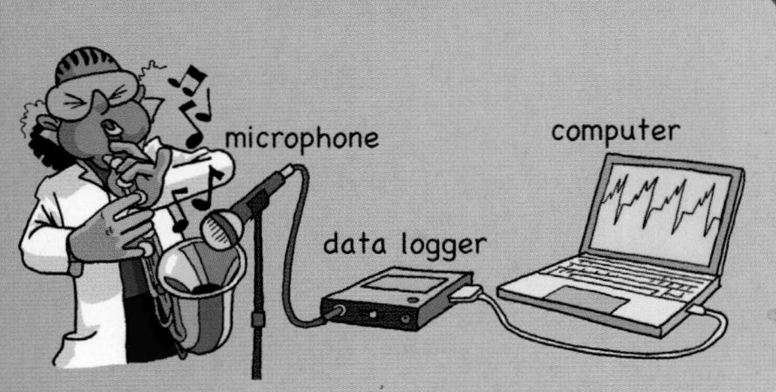

Summary Questions

1 Copy the sentences, choosing the correct word from each pair.

a) A louder/softer note has a greater amplitude/frequency.

b) A note with a higher/lower pitch has a greater amplitude/ frequency.

2 a) In the diagram below, which trace (A or B) has the greater amplitude? Which is louder?

b) Which trace (C or D) has the greater frequency? What difference would you notice if you heard these sounds?

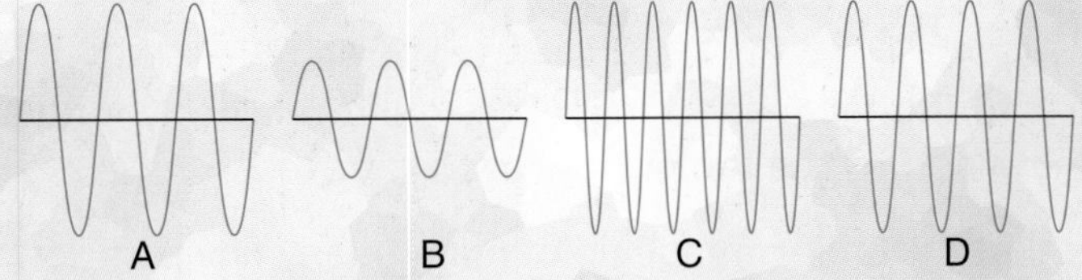

3 Different musical instruments can play the same note, but they sound different. Their oscilloscope traces are different. Find out what is the same for different instruments and what is different.

Link up to...

MUSIC

You may have used synthesisers in Music lessons. Connect the output of a synthesiser to an oscilloscope to see the patterns produced by different instruments.

KEY WORDS

signal generator
pitch
loudspeaker
oscilloscope
amplitude
frequency

How Sound Travels

- What materials will sound travel through?
- Will sound travel through a vacuum?
- How does sound travel?

Elephants are believed to communicate over long distances by sending low frequency vibrations through the ground – they detect them with their legs

Travelling through matter

When someone speaks to you, the sound of their voice travels through the air to your ears. This shows that sound can travel through a gas.

Tap yourself on the head. You hear the sound, which has travelled through the bone of your skull. Sound can travel through solids.

Many underwater creatures such as whales and dolphins communicate by sound through water. Sound can travel through liquids.

Sound always needs a material to travel through – it can be solid, liquid or gas. However it cannot travel through empty space (a vacuum).

activity

In space, no-one can hear the doorbell

A bell rings inside the glass jar. Its sound travels through air and glass to reach your ears.

- Pump out the air and the sound fades away.
- How can you tell that the bell is still ringing?

Safety: Use a safety screen.

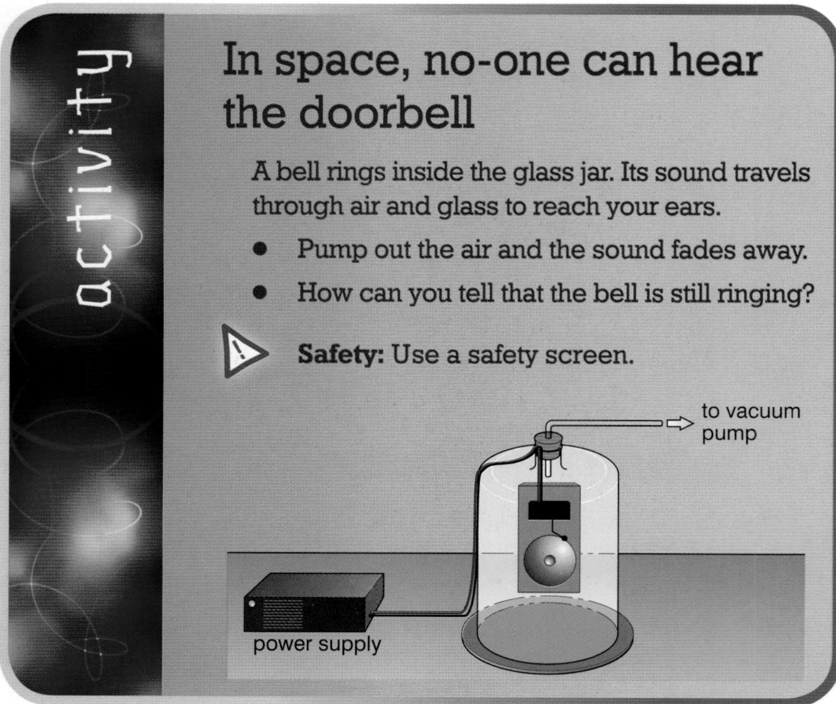

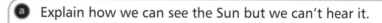

a Explain how we can see the Sun but we can't hear it.

Vibrating particles

Put your ear to the table and tap it. The sound reaches your ear through the wood of the table. How does it travel?

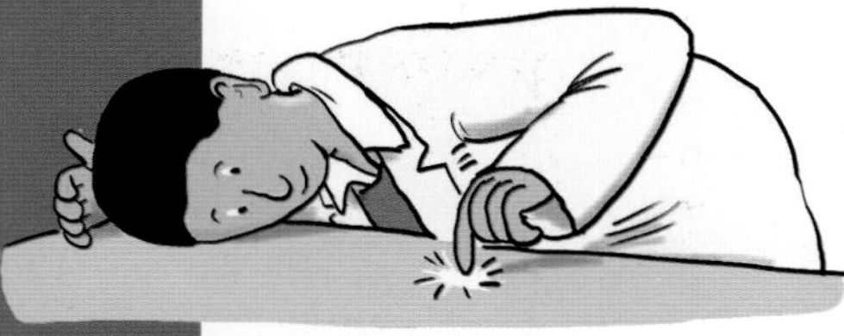

Remember that the table is made up of many, many tiny particles, closely packed together. When you tap the table, your finger makes the particles next to it **vibrate**. These particles push on their neighbours, which push on their neighbours, and so on. So the vibration is passed from one particle to the next, until it reaches your ear.

Sound waves

The particles don't travel all the way from your finger to your ear. They have fixed positions and they simply **oscillate** back and forth. Only the vibration travels from finger to ear.

Sound travels more slowly through air. The particles are farther apart than in a solid, so they don't collide with each other so often and so the vibration travels more slowly. Also, there are no particles in a vacuum to oscillate and pass on the vibration.

Vibrations travelling like this are called **sound waves**. They are like the waves you get at some swimming pools. The wave machine is at the deep end; it pushes the water back and forth and waves travel along the surface of the water to the shallow end. Water doesn't travel from one end of the pool to the other, but the waves do.

Did You Know?

The Spanish navy had to stop using sonar systems for detecting submarines because the underwater sounds were harming whales and dolphins.

Stretch Yourself

Use what you know about light and sound to explain why we see lightning before we hear thunder.

b Here is an alternative theory of how sound travels. 'When someone speaks to you, air travels from their mouth, carrying the sound with it.' How would you show that this theory is incorrect?

activity

Sound and solids

- Try some experiments to test which solid materials sound will pass through.
- Does sound travel better through some materials than others?
- Evaluate your investigation.

Summary Questions

1. What two words on this page both mean 'move back and forth'?
2. How could you show that sound can travel along a metal rod?
3. Put these in order, from slowest to fastest:

   a snail **sound waves in air** **a motorcyclist**
   light **sound waves in steel**

4. You may have used a 'Slinky' spring as a toy. Stretch a Slinky spring along the bench – ask a partner to hold the far end. Move your end back and forth gently. What do you observe? In what ways is this like a sound wave?
5. Do some research. Find out the speed of sound in different materials. Include solids, liquids and gases. Does sound always travel faster in solids than in liquids?

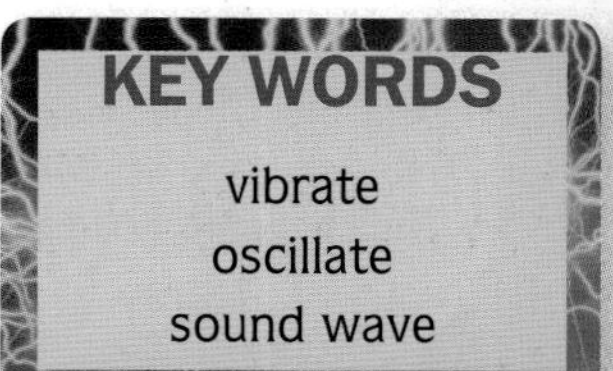

KEY WORDS

vibrate
oscillate
sound wave

Noise Annoys

P2.9

- What sounds can we hear?
- What is the difference between sound and noise?

High and low

What sounds can we hear? You can use a signal generator and loudspeaker to find out.

Turn the frequency down and down until the sound disappears. This usually happens at about 20 Hz (20 **hertz** – that's 20 vibrations per second). Then turn the frequency up and up until the sound is too high-pitched to hear. This happens at about 20 kHz (20 kilohertz, or 20 000 hertz).

This is the normal range of hearing for a child. Adults are usually unable to hear such high notes. However if you are a dog or a cat, you can probably hear much higher notes than your owner.

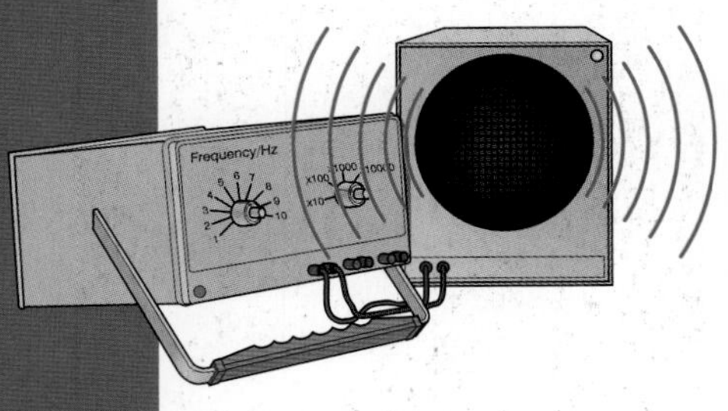

activity

Hearing test

- Carry out a test to determine your range of hearing.
- Then try another test to find out which member of the class can hear the faintest sounds.

a You can buy a dog whistle which makes a sound whose frequency is too high for people to hear. How could you show that it really does produce a sound?

In your ear

outer ear
middle ear
inner ear
to brain
ear canal
nerve
eardrum
cochlea
3 small bones

The human ear is a very sensitive organ. When you hear a sound, only a very small amount of energy enters the ear.

- When sound waves enter your ear, they press on the eardrum and make it vibrate.
- The three small bones make the vibrations bigger and pass them on to the cochlea.
- The cochlea sends a message along the nerve to your brain.

Such a sensitive organ is easily damaged. As you get older, it gradually deteriorates, making it harder to hear high-pitched or faint sounds.

Shouldn't we turn the volume down a bit?
What did you say?

Next time you...

... listen to loud music, think of those three delicate bones being bashed about in your middle ear.

Doing damage

The loudness of sounds is measured using a **sound-level meter**. This gives readings on the **decibel** scale.

- 0 decibels – the faintest sound you can hear
- 50 decibels – quiet conversation
- 80 decibels – a door slamming
- 110 decibels – a pneumatic drill
- 130 decibels – the threshold of pain

You can see that the loudest sounds are painful. Doctors have warned that many young people are damaging their hearing by listening to loud music through headphones.

This council official is checking noise levels on a by-pass

b Where would you put these on the decibel scale?
a ticking watch   loud conversation   a loud car horn

Noise

Loud music can be enjoyable, but it may annoy other people. Any unwanted sound like this is called **noise**. Noise in the environment is called **noise pollution** and it can make you ill. People who live close to an airport, for example, may suffer from noise pollution. In the long run, this can weaken their ability to fight off diseases.

People who live in noisy places may have **sound insulation** materials fitted to their homes to reduce noise coming in.

This aircraft is about to land at Birmingham airport

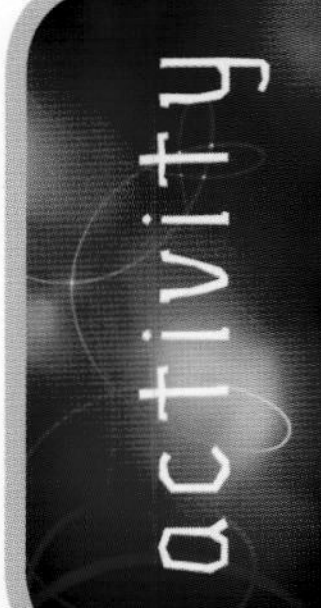

Keep that noise down!

- Try out a sound-level meter.
- How could it help to test sound insulation materials?

Engineers at Cambridge University have designed a 'silent aircraft', but manufacturers say that it is too expensive to build.

Summary Questions

1. Give a two-word phrase which means 'noise'.
2. Child A can hear sounds from 20 Hz to 15 kHz; child B can hear sounds from 20 Hz to 18 kHz. Which has the greater range of hearing? Why might the other child have a poorer range?
3. How could you use a sound-level meter and a data-logger to measure and record noise levels around your school or home?
4. Double-glazing can help to reduce noise. Find out other ways in which traffic noise can be reduced.

KEY WORDS

hertz
sound-level meter
decibel
noise
noise pollution
sound insulation

know your stuff

▼ Question 1 (level 5)

Jade was investigating how a beaker of hot water cooled down. The drawing shows her apparatus.

a Name the instrument she used to measure the temperature of the water. [1]

b Heat energy escaped from the beaker of water in several different ways. The table shows some of these. Copy and complete the table by filling in the last column. Choose from:

conduction convection insulation radiation

How the heat energy escaped	Name for this
Heat energy passed through the glass of the beaker and spread into the top of the table.	
Heat energy was carried away by warm air rising above the beaker.	

[2]

c After a while, Jade measured three temperatures:

temperature of water in beaker = 20°C

temperature of air near beaker = 20°C

temperature of table under beaker = 20°C

Predict whether the water would continue to get colder. Give a reason to support your answer. [2]

▼ Question 2 (level 5)

In Pete's room, there is an electric heater. Heat energy is carried around the room by convection currents. Air close to the heater is heated.

a How does the temperature of this air change? [1]

b How does the density of this air change? [1]

When air is heated, the speed and the separation of the particles of the air change:

c How does their speed change? [1]

d How does their separation change? [1]

Warm air above the heater rises:

e What happens as the warm air rises? [1]

▼ Question 3 (level 6)

Jude was investigating the different notes produced by his flute. He played it next to a microphone which was connected to an oscilloscope. The diagrams show some of the traces he observed.

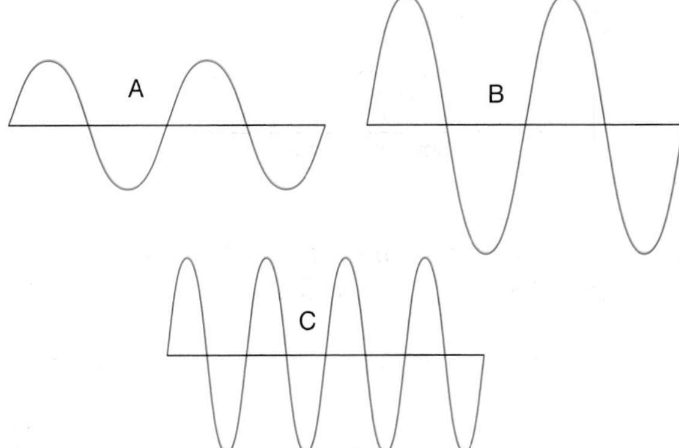

a Which diagram (A, B or C) represents a loud, high-pitched note? [1]

b Which diagram represents a loud, low-pitched note? [1]

c Copy diagram A. Add to it a trace which has the same frequency but a greater amplitude. [1]

d How would this note sound, compared with the note represented by trace A? [2]

How Science Works

▼ Question 1 (level 5)

Mel was investigating the freezing of seawater. She put a plastic beaker of seawater into a freezer, together with a temperature sensor connected to a computer to record the temperature of the water.

a An electronic temperature sensor is better than a liquid-in-glass thermometer in this experiment. Which two of the following are correct reasons for this? Give the letters. [2]

- **A** An electronic sensor records the temperature many times each minute.
- **B** A thermometer may give incorrect readings.
- **C** You would have to open the freezer to read the thermometer.
- **D** A thermometer cannot read below 0°C.

The graph shows how the temperature of the water changed:

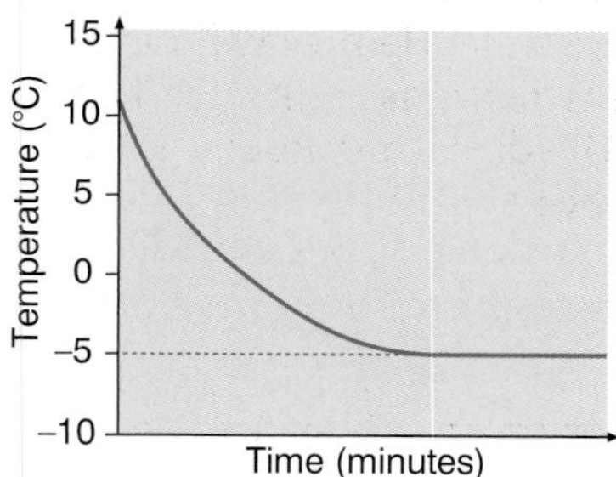

From the graph:

b What was the temperature of the water when Mel put it in the freezer? [1]

c At what temperature did the water freeze? [1]

The seawater froze at a different temperature to pure water:

d At what temperature does pure water freeze? [1]

e Mel repeated the experiment with pure water instead of seawater. Copy the graph and add to it to show how you would expect her results to appear. [2]

▼ Question 2 (level 6)

Tomek was investigating heat insulation. He heated a metal block by placing it in a beaker of boiling water. Then he removed it from the water, dried it and wrapped it in cotton wool. He used a thermometer to measure the temperature of the block as it cooled.

The table shows his results.

Time from start (minutes)	Temperature (°C)
0	76
1	68
2	62
3	56
4	51
5	46

a Draw a graph to show Tomek's results. [2]

b What was the temperature of the block when Tomek started measuring its temperature? [1]

c By how many degrees had the temperature of the block fallen after 5 minutes? [1]

Tomek wanted to show that the block cooled more slowly when it was wrapped in cotton wool than when it was not insulated.

d What other measurements should he make in order to show this? [2]

e On your graph, show the results you would expect Tomek to obtain if he used a *thinner* layer of cotton wool insulation. [1]

▼ Question 3 (level 7)

Jo and Edgar were investigating different materials, to see which was the best absorber of sound. Jo put squares of material over her ears, and Edgar spoke loudly nearby. Jo judged which material let through the least sound.

a List *three* things which Jo and Edgar should keep the same for this to be a fair test. [3]

b Describe how you could use a signal generator, a loudspeaker and a sound-level meter to investigate sound absorption. [2]

c An experiment using these scientific instruments would give more reliable results. Explain why the results would be more reliable. [3]

How Science Works

▸▸ How can we carry out project work effectively?

Carrying out Project Work

Collecting evidence

As you know, scientific enquiries are all about finding the answers to scientific questions.

In Fusion 1 (pages 152–161) you looked at some of the approaches you can use.

Here is a list to remind you:

- observing and exploring
- researching – using secondary sources
- classifying and identifying
- fair testing – controlling variables
- pattern seeking – surveys and correlation
- using models
- using and evaluating a technique or design.

We often call an extended enquiry 'a project'.

In science we collect evidence to answer the question we are investigating.

How we gather the evidence will depend on the question. In a project, it may well involve more than one of the ways listed above. So project work will take several lessons and/or homeworks to complete.

Group work

You might also be working as part of a group in project work.

a What are the benefits of working as a group in project work?

When you start on a project you must be clear about the question you are investigating. As a group you should discuss this together and sort out the different aspects of the problem. Then you can divide the work up between the members of the group.

Presenting your findings

It is a good idea to try to draw out a plan for the project at this stage. Perhaps use a flow chart.

Try to have a clear idea of what the finished project will look like. Ask yourselves 'What is the best way to share our findings with other people in the class or school?'

b What way do you like to receive information from other groups after project work? Why?

There are a variety of ways to present your findings to other people

Carrying out research

After your plan, many projects will involve research to find out what other people have found out already.

Before you start any research write down what you know already. Then record what you want to find out.

Decide where to look for the information. Will you use books, videos, CD-Roms or the Internet?

Once you have found some useful information, summarise it in your own words; do not copy or cut-and-paste directly. Also quote the source of your information. Think about whether you can trust it or not.

Interesting projects

The best questions to investigate are ones you are interested in yourself.

- Work in a group of four.
- Come up with a selection of scientific questions that you would like to investigate.
- Your teacher will use the list to see what interests you and to advise you what is possible.

Summary Questions

1. Think about the project you have enjoyed most in your time at school. What was the project and what made it so enjoyable?
2. Write a checklist for a group to evaluate their project when it is finished.

Communicating New Science

▸▸ How do scientists share their findings with the world?

Publishing results

When scientists carry out research and make new findings, they need to tell other scientists about their work. In this way, science makes progress. It carries on developing our understanding of the world.

A system called **'peer review'** is used before work gets published in scientific journals. A scientist will write a paper and send it to a journal. The journal will organise other scientists to look over (review) the paper before publication. It's a bit like you asking a classmate to check your work before you hand it in.

These reviewers raise any queries with the author of the paper. If they are happy with the author's replies and changes, the paper will be published. Then all scientists can read about the new work. However now and again the system does not stop 'bad science' getting through.

a Why are scientific papers reviewed before publication?

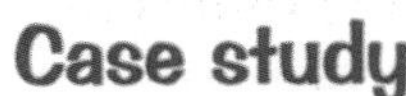

Case study

In 2001 Dr Jan Hendrik Schon was a rising star in one of the world's best research establishments. Many important inventions were made there. For example, in 1947 the transistor was developed. Now billions of these are made for our computers every week. Other inventions made there include mobile phones, lasers, communication satellites and internet systems.

Dr Schon worked in 'nanotechnology' – making incredibly tiny machines. He was trying to make the smallest computer components ever. His team claimed to discover single molecules that could act as transistors.

Some of the breakthroughs made at Bell Laboratories where Dr. Schon worked

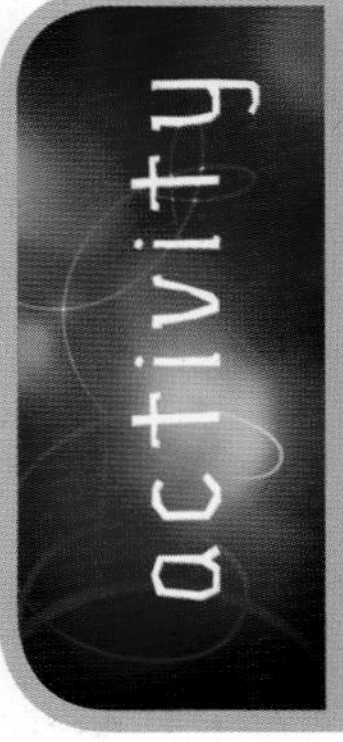

Great inventions

- Carry out some research into the work carried out since 1925 at the famous Bell Laboratories.
- Choose one example to write a two-minute item for a radio programme about 'Milestones in the last century of science and technology'.

In just two years he produced 80 publications. However, some scientists started to question some of the data in his publications.

They noticed identical tables of results in different experiments. Even readings that were meant to be random were the same. Other scientists tried out the same experiments themselves. They failed to reproduce Schon's data. So an inquiry took place into Schon's work.

b Why did scientists start to doubt Schon's work?

The inquiry found that Schon had made up experimental data at least 16 times between 1998 and 2001. They accused him of:

- removing data that disagreed with his predictions and
- making up data from mathematical equations, pretending they were really experimental results.

Schon disagreed with some of their findings. He claimed all his publications were based on experimental observations. However, he did admit making 'various mistakes in my scientific work, which I deeply regret'. Despite his objections, Bell Laboratories quickly fired Schon.

This has raised questions about the 'peer review' system. The scientists who review the papers might not query the results of work from large institutions or from famous scientists.

Reviewers claim that their job is to check the validity of conclusions drawn from data. They do not check the data themselves. They trust the scientists to record their actual results. So the system remains open to abuse from dishonest scientists.

To combat this, editors want reviewers to question any data from single or a small number of experiments – whatever their source!

Summary Questions

1. How do scientific journals try to make sure the papers they publish are 'good science'?
2. What kind of things might influence a scientist to publish false data?
3. What do you think can be done to make it harder for 'bad science' to get published? Point out any difficulties your ideas might cause.
4. How do you try to make your own experimental data more reliable?

KEY WORDS

peer review

The Application of Science

- How are discoveries made in science?
- How can we apply new discoveries to make useful products?

The start of the plastics industry

Have you ever wondered what life would be like without plastics?

A chemist named Leo Baekeland really started our 'plastic society'. Leo was a chemist with a sharp eye for a business opportunity. Although he was born in Belgium in 1863, he made his fortune in America.

He investigated chemicals you could get from coal tar and wood alcohol. 30 years earlier a German dye chemist had discovered a thick, gooey substance while experimenting. He thought it was just a nuisance. However Leo could see its potential for use as a new varnish.

If Leo heated up the gooey liquid it turned even thicker. Also if he did the reaction under pressure it made a hard, translucent solid. He could mould this into any shape he liked. So in 1907 he had made the first synthetic plastic! It had taken him 3 years and thousands of failed experiments. However, his perseverance paid off in the end.

Leo knew how to make the most of his discovery. He patented his new plastic and gave it the trade name 'Bakelite'. He formed a company to make and sell the plastic.

People were quickly using the new plastic, for everything from buttons to telephones. Leo could make his plastic in a variety of colours. It even became trendy to wear Bakelite jewellery. The plastics revolution had started!

The casing of this old radio is made from Bakelite

a Make a list of all the plastic things you have used so far today.

activity

Plastics research

Do some research to find out about either:

- the problems of plastic waste and how we might solve it, or
- the life and work of Wallace Carothers.

Surprising discoveries

In 1938 Dr Roy Plunkett was experimenting with substances to keep fridges cold. One day he found a waxy solid blocking a gas cylinder. This accidental discovery turned out to be a new **polymer**. A polymer is a very long molecule made by joining lots of smaller molecules. This polymer was called polytetrafluoroethene or PTFE.

It was sold as Teflon. This special polymer has special 'non-stick' properties. It is also very unreactive. This has led to a wide variety of uses, from non-stick pans to rocket cones.

Non-stick pan

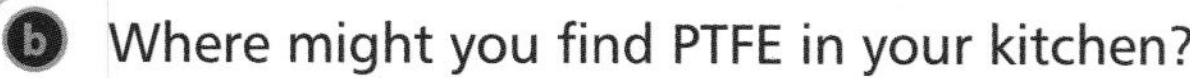

b Where might you find PTFE in your kitchen?

The racing 'leathers' of this rider are in fact synthetic and contain Kevlar to protect against abrasion in case of an accident

In the 1960s the search was on for a new polymer to make lighter weight tyres. This would help save fuel. Stephanie Kwolek and her team worked on the problem. One day the chemicals she mixed formed a milky liquid, unlike the clear liquid she was expecting. However she didn't just throw the liquid away and start again.

She used the liquid to make a new material and sent her discovery to the test lab. This stuff was incredible! It was nine times stronger than a similar mass of steel, but was only half the density of fibreglass. Eventually it was marketed in 1971 as Kevlar.

Its strength, low density and heat resistance led to its use in bullet-proof vests, aeroplanes, motorcycle 'leathers' and tennis rackets.

c How has Kevlar helped save many lives?

Sheets of Kevlar fibre are compressed together to make body armour

Summary Questions

1. Describe how two creative scientists have helped to change people's lives by their work in the plastics industry.
2. Most plastics nowadays are made from products we get from crude oil – a fossil fuel.

   How do you think the plastics industry will develop in the next 50 years?

KEY WORDS

polymer

Glossary

abdomen the lower part of the body, below the diaphragm, containing stomach, intestines, kidney, liver and, in women, the uterus

absorb when a ray of light is taken in when it strikes a surface

accurate the results are close to the truth

acid rain rain with a pH lower than that of natural rain water

aerobic respiration respiration using oxygen

alveolus air sac in the lungs where gases pass between the air and the blood

amino acid molecules that are joined together to make protein molecules

amplitude the height of a wave

amylase enzyme that breaks down starch into sugar

anaerobic respiration respiration without oxygen

artery blood vessel that carries blood away from the heart

-ate a compound that contains a non-metal and oxygen

atom the smallest particle that can exist on its own

axis the imaginary line through the North and South Poles, about which the Earth turns

balanced equation the same number and type of atom on each side of a symbol equation

bile substance produced by the liver which helps digestion, especially of fat

bioaccumulation a process in which substances such as metals or pesticides, from food, build up in the body then are passed on to the next animal in a food chain

biological weathering when a plant or animal breaks rock into smaller pieces

boiling point the temperature at which a substance turns from a liquid to a gas or from a gas to a liquid

branching a type of key

branching bronchi tube that carries air from the trachea into the lung

bronchiole smaller branch of the bronchi

capillaries smallest blood vessels, with a thin wall through which substances diffuse in and out of the blood

carbohydrate a food type used for energy; including sugars and starch

carbonic acid the natural acid found in all rain water, formed as carbon dioxide dissolves in water

cardiac muscle special muscle that the heart is made of

carnivore an animal that eats meat

Celsius scale a scale of temperature

cementation when rock particles are stuck together by minerals, to make sedimentary rock

chemical bond the force between atoms, holding them together

chemical digestion breaking down of large food molecules into smaller ones, which can be absorbed

chemical weathering when a chemical reacts with a rock to weaken the rock and make new substances

chlorophyll the green substance in plants which collects light energy

chloroplast a structure in a cell which contains chlorophyll and is the site of photosynthesis

cilia small hair-like structures on the cells lining the lungs, which brush debris from the lungs

combining power how many chemical bonds an atom can make

compaction when sediment is punched together

competition this takes place when two animals or plants need the same resource

compound more than one type of atom, chemically joined

consumer an organism that gets its food by eating other living things

contract get smaller

convection current when heat energy is carried by a moving liquid or gas

crater dent in the Moon's surface, caused by rocks from space

decibel a unit used to measure the loudness of sound

deciduous trees that lose their leaves in winter

decomposer an organism that breaks down dead animal and plant remains

density density $= \frac{\textbf{mass}}{\textbf{volume}}$

deoxygenated blood that has had oxygen removed from it

deposited when sediment is dropped from water or wind

diaphragm a sheet of muscle between the abdomen and thorax which helps us to breathe in and out

dichotomous a type of key

element only one type of atom and it is listed on the Periodic Table

emphysema a disease where the walls of the alveoli break down, reducing the surface area of the lungs

energy a variable, measured in joules (J)

enzyme a protein molecule that controls a chemical reaction in living organisms (including digestion of food)

erosion weathering and transportation

error a result that you can tell is wrong

excretion removal of waste chemicals, such as urea and carbon dioxide, that are produced by the body

exhale breathe out

expand get bigger

faeces indigestible remains of food, mixed with bacteria, commonly known as 'poo'

fat a food type which gives large amounts of energy and is essential in small amounts for making cell membranes

filter a piece of transparent, coloured material

food chain a diagram that shows the feeding relationships between animals and plants

food web a series of linked food chains in a habitat

formula shows the number and type of atoms in a molecule and other compounds

freeze–thaw a type of physical weathering, where the repeated freezing and thawing of water in rocks causes them to break

frequency the number of waves or vibrations per second

gas giant a large planet made of frozen gases, e.g. the outer planets of our solar system.

gaseous exchange passage of carbon dioxide from the blood to the air in the lungs, and oxygen from the air in the lungs to the blood

glycogen a compound made of glucose molecules joined together used as a store of energy in animals

guard cell the cells around the stomata in a leaf, which make the stomata open and close

habitat the place where an animal or plant lives and reproduces

haemoglobin a protein, containing iron, used to carry oxygen in the blood

heart the organ that pumps blood around the body

heat energy (heat) energy spreading out from an object which is hotter than its surroundings

hemisphere half of the Earth's sphere

herbivore an animal that eats plants

hertz the unit of frequency; 1 Hz = 1 vibration per second

hibernate slowing down body activity to a minimum and spending winter conditions in a dormant state

homeostasis keeping body conditions, such as temperature, water content and salt concentration, constant

-ide a compound that contains a non-metal

idea a suggestion explaining something, without any evidence to prove your points

identification key a series of questions that are used to help identify an organism

igneous rocks made from cooled magma

image a picture of something, formed by light

incident describes a ray of light falling on a surface

infra-red radiation invisible rays of heat energy spreading out from an object that is hotter than its surroundings

inhale breathe in

intercostal muscles muscles between the ribs which are used in breathing in and out

key a tool to help you identify and classify things

larynx the' voice box', used in making sounds

lava magma above the Earth's crust

lens a piece of glass shaped so that it refracts rays of light so that they come together

limiting factor a factor, such as temperature, that controls a chemical reaction

lipase an enzyme that breaks down fats

loudspeaker a device that turns electricity into sound, so that we can hear it

lunar eclipse when the Earth blocks the Sun's rays so that the Moon is in darkness

magma molten rock below the Earth's crust

material a word that scientists use to descibe what objects are made from

melting point the temperature at which a substance turns from a solid to a liquid or a liquid to a solid

metal a chemical that is shiny, malleable and a conductor; they often have high melting and boiling points and are sonorous

metamorphic rocks that have been changed by heat and/or pressure

migrate to move to another, more favourable area on a seasonal basis

mineral (biology) an element which is needed by animals and/or plants to grow

mineral (chemistry) an element or compound found in rocks

molecule more than one atom, chemically joined

mucus sticky substance produced by cells lining the respiratory surfaces; it traps bacteria and dirt

neutralisation a chemical reaction between a base (or alkali) and an acid; this makes a metal salt and water

nitrogen oxide a gas made in the engine of a car that can cause acid rain

nocturnal describes an animal which sleeps in the day and is active at night

noise unwanted sound

noise pollution unwanted sound in the environment

non-metal a chemical that is dull, brittle and an insulator; they often have low melting and boiling points

normal the line at right angles to a surface at the point where a ray of light strikes the surface

omnivore an animal that eats both plants and animals

onion-skin weathering a type of physical weathering, where the repeated heating and cooling of the rock causes its outer layer to peel off

oscillate to move back and forth

oscilloscope an electronic instrument used to show sound waves on a screen

oxidation a chemical reaction where oxygen is added

oxygen a gas that makes up about 20% of the air

oxygen debt oxygen which has to be used to break down lactic acid produced in anaerobic respiration

oxygenated blood containing oxygen

palisade cell a cell below the upper surface of a leaf where most photosynthesis takes place

parasite an animal or plant that lives on another animal, causing it some harm

particle model a way of thinking of matter as being made up of particles

peer review the checking of a scientific paper by fellow scientists before it gets published

penumbra region of partial shadow in an eclipse

Periodic Table a list of all the elements that make up the universe

peristalsis contractions of the muscles of the intestines, pushing food through the gut

pesticide a chemical that kills animals which are a pest, e.g. in gardens and farms

phases a stage in the changing appearance of the Moon, as seen from Earth

phlogiston a theory to explain combustion that has now been disproved

photosynthesis the process by which plants use light energy to convert carbon dioxide and water into glucose

physical digestion breaking down large lumps of food to smaller ones, by chewing and by contractions of the gut wall

physical factor a factor such as temperature or light which affects the organisms living in a habitat

physical weathering the breaking down of rocks due to wind, water, other rocks hitting them or the Sun

pitch how high or low a note is

pitfall trap a small trap set in the ground to catch small animals

planet a large, solid object in orbit around the Sun or another star

plasma the liquid part of the blood

platelets small cell-like fragments in the blood which help in clotting

pollution changes to the environment by humans because of noise, chemicals and heat

polymer very long molecule made by joining lots of small molecules together

pooter a suction device for catching small animals

predator an animal that catches and eats other animals

pressure a variable measured in Pascal (Pa)

prey an animal that is caught and eaten by another animal

prism a glass block with a regular shape

producer a green plant that makes food from carbon dioxide and glucose using light energy

products the end chemicals in a reaction

properties a description of how a chemical will look and behave

protease an enzyme that breaks down proteins

protein a food type used for growth and repair and found in meat, eggs, cheese and beans

pyramid of numbers a diagram that shows how many organisms there are at each stage of a food chain

quadrat a square used for collecting information about the number of plants in a habitat

radiation energy spreading out from its source

range the difference between the smallest and largest values of a quantity

reactants starting chemicals in a reaction

recycle collection and processing of used materials to make new things

red blood cells cells which carry oxygen in the blood

reflect when a ray of light bounces off a surface

reflected describes a ray of light which has bounced off a surface

reflex action rapid, automatic response by the nervous system

refraction the bending of light when it travels from one material to another

reliable every time the experiment is repeated, the results are almost the same, making them more trustworthy

rock cycle theory to explain how rocks are recycled on Earth

rocky planet a small planet made of rocky material, e.g. the four planets closest to the Sun

saliva liquid secreted into the mouth which contains amylase and also lubricates food so that it easily slides down the oesophagus

sampling a method of collecting data about the number of animals and plants in a habitat

scatter when a ray of light reflects in many different directions

scavenger an animal that eats animals which are already dead

season a division of the year

sedimentary rocks made from sediment that has been compacted and cemented together

sensitive able to detect small differences or changes

signal generator an electronic device used with a loudspeaker to produce sounds of different frequencies

solar eclipse when the Moon blocks the Sun's rays so that part of the Earth is in darkness

sound insulation material used to absorb sound

sound wave how we picture sound travelling through a material

sound-level meter a meter used to measure the loudness of sound

spectrum all the colours of light, spread out in order

speed of light the speed at which light travels in empty space; about 300 000 km/s

star a glowing mass of hot gas, e.g. the Sun

starch a type of carbohydrate found in potatoes, bread, rice and pasta

stimulus something that is detected by the senses and triggers a nerve impulse, leading to a response

stomata the small pores in a leaf that let in carbon dioxide and let out water and oxygen

sugar type of carbohydrate, found in sweet foods

sweep net a net used to catch small animals living in long grass

sulfur dioxide a gas made in the engine of a car that can cause acid rain

supernova an explosion of a massive star towards the end of its life

symbol a short, unique code for every element

symbol equation shows the symbols and formulas for the starting and ending substances in a chemical reaction

synthesis a type of chemical reaction, where a compound is made from its elements

temperature a measure of how hot something is

theory an explanation for an observation with facts to help prove your ideas

thermal conductor a material through which heat passes easily

thermal decomposition a chemical reaction, where heat is used to break down a substance to simpler chemicals

thermal insulator a material through which heat passes only very slowly

thermometer an instrument for measuring temperature

thorax the upper part of the body containing the lungs and heart, sparated from the abdomen by the diaphragm

tissue fluid fluid formed from plasma 'leaking' from the blood and surrounding the cells of the body

trachea the windpipe, linking the larynx to the bronchi

transmit when a ray of light passes through a material

transportation moving from one place to another

tree beating a method of catching small animals by hitting a tree or shrub so they fall out onto a sheet on the ground

umbra region of full shadow in an eclipse

upthrust the upward force on an object in a liquid or a gas

urea a toxic chemical produced from waste protein

variable a quantity that can be changed, controlled or measured in an experiment

vein blood vessel that carries blood towards the heart

vibrate to move back and forth

villi small projections on the lining of the intestine which increase the surface area for absorbing food

vitamin a compound needed in very small quantities in the diet, essential for health

weathering the breaking down of rocks

white blood cells blood cells involved in fighting diseases

word equation shows the names of the starting and ending substances in a chemical reaction

Index

Acknowledgements

Alamy 91, 112, 114.1, 120, 124, 130.2, 132, 136, 147.1, 147.2; **Andrew Lambert** 72, 74; **Bruce Coleman** 95.2; **Corbis** 44.3, 47.2, 112; **Fotolia** 35, 37, 39.1, 39.2, 42.2, 44.1, 44.2, 44.4, 44.5, 44.6, 70, 86.1, 86.2, 88.1, 88.3, 88.2, 101, 152.1, 152.2, 152.3, 154, 155.1, 155.2; **GeoScience Features Picture Library** 84.1, 84.2, 84.3; **Getty** 80, 83, 89, 90.2; **GreenGate** 54; **iStock** 21, 32, 34.1, 42.3, 42.4, 45.2, 45.3, 45.4, 45.5, 45.6, 58.1; **Mark Boulton** 133; **M Chillmaid** 117; **Nasa** 121.2, 123; **Nature picture Library** 106, 140.2, 144; **NHPA** 130.1; **OSF** 58.2; **Photolibrary** 42.1, 43.1, 43.2, 45.1, 46.1, 47.1, 47.3, 47.4; **Photo researchers** 76.2, 107, 121.1; **Science Enhancement Programme** 131; **Science Photo Library** 4, 13, 17.1, 17.2, 18, 19, 34.2, 36, 46.2, 46.3, 46.4, 57, 60.1, 60.2, 60.3, 62, 64, 76.1, 76.3, 77, 82, 84.4, 90.1, 95, 98.1, 98.2, 99.1, 100, 102.1, 102.2, 103, 104, 108, 114.2, 116, 118, 119, 122.1, 126, 127, 131, 138, 140.1; **Topfoto** 66

Picture research by GreenGate Publishing

Every effort has been made to trace all the copyright holders, but if any have been overlooked the publisher will be pleased to make the necessary arrangements at the first opportunity.

Notes